Surgical Anatomy of the Face

Surgical Anatomy of the Face

Wayne F. Larrabee, Jr., M.D., F.A.C.S.
Clinical Professor
Department of Otolaryngology–Head and Neck Surgery
University of Washington School of Medicine
Seattle, Washington

Kathleen H. Makielski, M.D.
Associate Professor
Department of Otolaryngology–Head and Neck Surgery
University of Washington School of Medicine
Chief, Otolaryngology–Head and Neck Surgery
Pacific Medical Center
Seattle, Washington

Illustrated by Kathleen H. Makielski, M.D.

Raven Press New York

Raven Press, Ltd., 1185 Avenue of the Americas, New York, New York 10036

Made in the United States of America

Library of Congress Cataloging-in-Publication Data

Larrabee, Wayne F.
Surgical anatomy of the face/Wayne F. Larrabee, Jr., Kathleen H. Makielski; illustrated by Kathleen H. Makielski.
p. cm.
Includes bibliographical references and index.
ISBN 0-88167-945-3
1. Face—Anatomy—Atlases. 2. Anatomy, Surgical and Topographical—Atlases. I. Makielski, Kathleen H. II. Title.
[DNLM: 1. Face—anatomy & histology—atlases. 2. Face—surgery—atlases. 3. Surgery, operative—methods—atlases. WE 17 L333s]
QM535.L35 1992
611'.92—dc20
DNLM/DLC
for Library of Congress 92-21808
CIP

9 8 7 6 5 4 3 2 1

My interest in facial anatomy was stimulated by my uncle, Raymond Truex, Ph.D. (1911–1980), whose books on neuroanatomy and head and neck anatomy set new standards of scholarship, accuracy, and design. I dedicate my efforts in *Surgical Anatomy of the Face* to my brother Gregory R. Larrabee (1949–1976), a student of anatomy, and to my wife Suzette.

Wayne F. Larrabee, Jr., M.D.

I dedicate this book to Professor Gerald P. Hodge, who has inspired my interest in medical illustration ever since I learned of his work when I was 11 years old. He is an exceptional artist and teacher, and has continued to foster my interest in scientific art even after I pursued a career in surgery.

Kathleen H. Makielski, M.D.

Contents

Contributing Authors

Craig L. Cupp, M.D. *Clinical Assistant Professor, Department of Otolaryngology–Head and Neck Surgery, University of Washington School of Medicine, Seattle, Washington 98195*

Douglas J. Kibblewhite, M.D., F.R.C.S.(C) *Clinical Instructor, Department of Otolaryngology–Head and Neck Surgery, University of British Columbia School of Medicine, Vancouver, B.C., Canada*

Brock Ridenour, M.D. *Clinical Instructor, Department of Otolaryngology–Head and Neck Surgery, University of Washington School of Medicine, Seattle, Washington 98195*

List of Figures

Chapter 9. Vascular Patterns of the Face

Chapter 10. Lymphatics of the Face

Chapter 11. Hair and Scalp

Chapter 12. Forehead and Brow

Chapter 13. Eyelids, Anterior Orbit, and Lacrimal System

Chapter 14. Nose

Chapter 15. Ears

Chapter 16. Cheeks and Neck

Chapter 17. Lips and Chin

Foreword

Dr. Wayne Larrabee and Dr. Kathleen Makielski have certainly obtained their objective of a presentation of the relationship of functional anatomy from a surgical point of view. They have accomplished this both conceptually and graphically with a clarity and specificity that deal with the "common sense" of surgical techniques. This position is embellished with a well-balanced style that includes the names of appropriate pioneers and innovators relating to specific surgical techniques. The selection of topics is naturally arbitrary, but inclusive enough to establish a starting point for any type of regional surgeon working in these areas of the head and the neck.

This book is balanced by a practical and artistic identification of the soft and hard geographic compositions of the head and the neck areas. It has additional regional orientation with an individual presentation of the highly specialized organ systems of this area. These chapters on regionality emphasize the importance of the sanctity of their physiologic and aesthetic roles.

The artwork deserves special attention because of its practicality and gracefulness with tasteful and colorful presentations of the surgical anatomy. The artistic use of the hard line and soft colors has added an illustrative and artistic dimension that is very helpful in understanding the anatomic process.

The synthesis of the anatomic design with the implied clinical relationship in a surgical program assists greatly in the organization of the text. The comfortable sequential flow of this specific scientific information with its unique emphasis on basic anatomy sets the stage appropriately for intelligent surgical intervention.

John Conley, M.D.

Preface

Our goal in this book is to present relevant facial anatomy from a surgical point of view. The many anatomy books available do not present structures as seen at the operating table and often lack perspective as to the importance of certain details. Conversely, the numerous articles and books that describe key anatomic points for a specific operation frequently lack graphic clarity and consistency. *Surgical Anatomy of the Face* has been designed to consolidate this information in an understandable manner. The subjects covered are selective and reflect the authors' judgments of relative importance. A selected list of references and readings, grouped by topic, appears at the end of this book.

In addition to creating aesthetically pleasing and accurate illustrations, we have worked to use color in a consistent fashion and to design innovative ways to show complex relationships. Clinical examples and cadaver dissections have been used freely to supplement the illustrations and demonstrate specific anatomic points. Finally, the illustrations, which depict individuals of both sexes and various ages and races, represent the wonderful diversity of our patients in the Pacific Northwest.

Wayne F. Larrabee, Jr., M.D., F.A.C.S.
Kathleen H. Makielski, M.D.

Acknowledgments

We wish to acknowledge the help of several individuals in preparing this atlas: Patti Peterson, for her considerable efforts and dedication in preparing the manuscript through its multitude of revisions; Ward Makielski, for his artistic advice and assistance with graphics; Douglas Wilson, Ph.D., for his support, encouragement, and editorial assistance; Carol Risan, for her administrative support; Daniel Graney, Ph.D., for reviewing the manuscript for anatomic accuracy; Craig Murakami, M.D., for his assistance with the dissections; Ernest A. Weymuller, Jr., M.D., for his departmental support.

We are also grateful to Raven Press and Kathey Alexander, Editor-in-Chief, for their support.

Surgical Anatomy of the Face

SECTION I

Facial Analysis

CHAPTER 1

Facial Contour Analysis

The anatomic structure of the face can be conceptualized as a tripartite composite of: (1) skin, (2) soft tissue (fat, muscle, and connective tissue), and (3) the hard tissue foundation (bone, teeth, and cartilage). The basic form of the face is determined by the hard tissues. The skin and the underlying tissues create a soft-tissue envelope.

Of particular importance to facial contour are the convex facial bones, e.g., the nasal bones, the supraorbital rims, the malar eminences, the mandible, and the hyoid bone. Facial plastic surgery is increasingly concerned with procedures on this bony framework. The relationship between changes made in the hard tissues and the ultimate soft-tissue position is complex, however. In an area with thin elastic skin (such as the nasal dorsum), a bony change may result in the same soft-tissue change; on the other hand, a change in the bony chin may result in less of a soft-tissue change in the thick overlying soft-tissue profile (Gallagher et al., 1984).

FACIAL PROPORTIONS

In the vertical direction, the facial thirds of Leonardo are simple but useful ratios (Fig. 1.1). The height of the upper lip is about half that of the lower lip and the chin. Horizontally, the width of the nose at its base should be approximately equal to the distance between the eyes (Fig. 1.2).

In analyzing the frontal view, one should also consider the overall shape of the face. A 3:4 ratio between the width and the height of the head is fairly typical, but there is wide variation. Faces can be classified as square, round, oval, or triangular. A square or a round face may suggest a somewhat wider and shorter nose than an oval or a triangular one. An oval face is usually considered most pleasing.

From the lateral view, the general shape of the facial profile is important in aesthetic surgery. The basic concept of facial convexity was well described by Woolnoth in 1865: "The general form and outline of all faces, especially as they are seen in profile, are of three orders—the straight, the convex, and the concave. The straight face is considered the handsomest." Gonzalez-Ulloa (1961) defined a straight face with his profileplasty; in his technique a line is dropped from the nasion perpendicular to the Frankfort horizontal and should touch the forehead, the lips, and the chin.

The relatively simple study of facial proportions presented thus far is adequate for evaluation of many patients. When a more detailed analysis of facial contour is required, the surgeon can proceed to hard-tissue cephalometrics, soft-tissue cephalometrics, or more complex three-dimensional methods.

CEPHALOMETRIC ANALYSIS

As aesthetic surgery involves more extensive work on the bony framework of the face, cephalometric analysis becomes more important. Multiple systems have been described; the majority have been developed by orthodontists to evaluate the relationship of the teeth to the surrounding bony and soft-tissue coverings. These systems can be quite useful, particularly in evaluating vertical facial proportions and the relationship of the maxilla and the mandible to the cranial base. In this chapter, we will present some of the basic terminology and a few of the more basic relationships to assist the readers in orientation. The systems selected by a surgeon will depend on the specific cases involved and on the desires and the training of the individuals. Measurements can be made from a lateral cephalometric film, direct soft-tissue facial measurements, photographs, and computed tomography (CT) or magnetic resonance imaging (MRI) scans.

Hard-Tissue Cephalometric Analysis

For hard-tissue cephalometrics, the American standard cephalometric arrangement is commonly used. Stabilization is achieved with a pair of ear rods in the external auditory meatus and with one rod resting on the infraorbital rim or the nasion. These rods also mark one of the major points, the upper margin of the external auditory canal. The line that connects this point to the orbitale is termed the Frankfort horizontal. Positioning the patient's head so that the Frankfort horizontal is parallel to the floor permits reproducible cephalometric measurement. As a simple alternative to the Frankfort horizontal, a true horizontal based on natural head position can be used for cephalometrics, photographic documentation, and other analysis. This natural head position is obtained by having the sitting patient look into his or her own eyes in a mirror. The position is easily achieved clinically and is quite reproducible. Some common cephalometric points used in hard-tissue analysis are seen in Figure 1.3.

One of the major difficulties in cephalometric analysis is the setting of normal reference standards. An early, widely accepted standard was the Downs analysis, based on a post–World War II study at the University of Illinois on 25 adolescent Caucasians with ideal dental occlusion. Since then, many studies have been published, including the Michigan Growth Study (Riola et al., 1974), and the Bolton study in Cleveland (Broadbent et al., 1975).

There are numerous cephalometric systems that use different linear distances between points or angles between lines to analyze faces. In essence, they are all attempting to relate the five major functional components of the face to each other, both horizontally and vertically. These components are (1) the cranium and the cranial base, (2) the skeletal maxilla (maxilla minus teeth and alveolar process), (3) the skeletal mandible (again minus teeth and alveolus), (4) the maxillary dentition, and (5) the mandibular dentition.

The Steiner analysis was the first hard-tissue cephalometric analysis to be widely accepted (Steiner, 1959). Because it is still commonly used, some of its components will be described to illustrate the approach taken by many of the systems (Fig. 1.4). The first measurement is of the SNA angle, which describes the anterior-posterior relationship of the maxilla to the cranial base. A large SNA angle implies maxillary protrusion; a small angle, maxillary recession. The second measurement is the SNB angle, which measures the relationship of the mandible to the skull base. A large SNB angle implies mandibular prominence; a small angle, mandibular recession. The ANB angle is the difference between the SNA and the SNB angles and represents the difference in relative positions between the maxilla and the mandible. The ANB angle, however, is also influenced by the vertical height of the face and the anterior-posterior placement of the nasion. Next in the Steiner analysis, the angles of the upper and lower incisors to the NA and the NB reference lines are measured, as well as the distance from the incisal edge to the lines. The relative position of the chin (pogonion) to the lower incisor is important; a relatively prominent incisor allows a more prominent chin and vice versa. Finally, the angle between the mandibular plane and the SN reference line gives a measure for the vertical proportions of the face. Typical values for these measurements are given in Figure 1.4.

Soft-Tissue Cephalometric Analysis

Although hard-tissue cephalometric analysis has been developed more thoroughly than soft-tissue analysis, the majority of facial reconstructive and cosmetic work still involves soft tissue. Additional analysis of soft-tissue contours is important in planning and evaluating this surgery.

Many of the points relating to the soft-tissue profile are defined similarly to the hard-tissue points. The soft-tissue Frankfort horizontal is defined as the horizontal line extending from the superior border of the external auditory canal to the inferior border of the infraorbital rim. Some of the major points used to define the soft-tissue facial profile are seen in Figure 1.5.

Representative soft-tissue cephalometric systems include those of Powell and Humphreys (1984), Peck and Peck (1970), and Holdaway (1983, 1984). Powell and Humphreys describe their "aesthetic triangle" (Fig. 1.6), which allows relative comparisons of the soft-tissue profile to be made from a variety of lateral facial representations, including photographs and soft-tissue films. Peck and Peck describe three angles that reflect vertical facial proportions (Fig. 1.7); these authors also relate the chin, the lip, and the nose to a unique orientation plane connecting the glabella and the pogonion (Fig. 1.8).

Holdaway describes a "harmony line," or H line, which extends from the pogonion to the most prominent part of the upper lip (Fig. 1.9). The soft-tissue facial line that runs from the soft-tissue nasion to the pogonion meets the H line to create the H angle. An average H angle is 10°; a larger angle relates to increasing soft-tissue profile convexity. This system relates the H line to many of the standard hard- and soft-tissue points noted previously.

In addition to the soft-tissue cephalometric measurements noted above, there are various graphic means to directly compare patient profiles to standards. Although these techniques lack precision, they do allow one to see where a facial pattern departs from normal form. Various pointing devices connected to microcomputers can be used to map key facial contour points in three dimensions (Fig. 1.10). A cephalometric system based on this technology should facilitate preoperative planning and postoperative analysis (Larrabee, 1988).

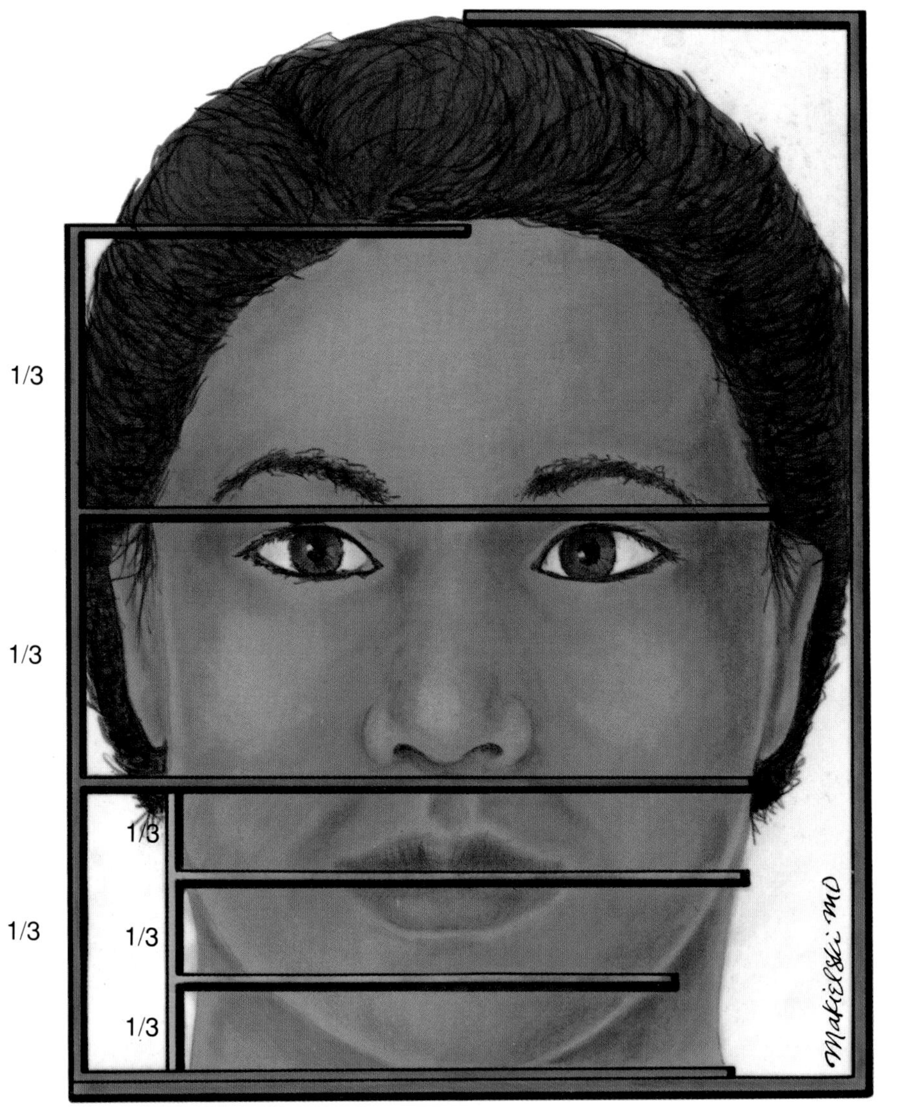

FIG. 1.1. Vertical facial proportions. Between the hairline and the chin, the face may be divided into thirds, usually just below the brows and at the base of the nose. The lower third of the face may be subdivided further into thirds, with the upper lip being half the height of the lower lip and the chin. The width to height ratio of the head is typically 3:4.

FIG. 1.2. Horizontal facial proportions. The width of the base of the nose is approximately equal to the distance between the eyes. The face is usually 5 eyewidths across.

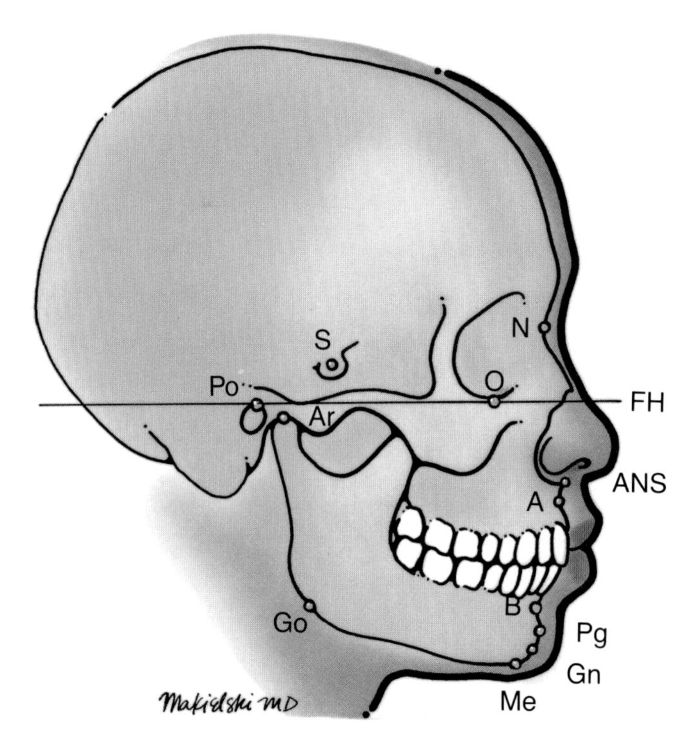

FIG. 1.3. Standard hard-tissue cephalometric points. The orbitale (**O**) is the most inferior point on the infraorbital rim. The nasion (**N**) represents the most anterior point at the nasofrontal suture. The center of the sella turcica is designated **S.** The tip of the anterior nasal spine is point **ANS.** The most retruded portion on the premaxilla between the nasal spine and the incisor is **A.** The deepest point of the mandibular bony profile is **B.** The most anterior point on the bony chin is the pogonion (**Pg**). The center of the inferior contour of the bony chin is the gnathion (**Gn**). The most inferior point on the bony chin is termed the menton (**Me**). The midpoint at the angle of the mandible is the gonion (**Go**). The point at the intersection of the posterior border of the mandibular ramus and the shadow of the zygomatic arch is **Ar.** The porion (**Po**) represents the midpoint of the upper part of the external auditory canal. The Frankfurt horizontal (**FH**) connects the superior aspect of the external auditory canal to the orbitale (**O**).

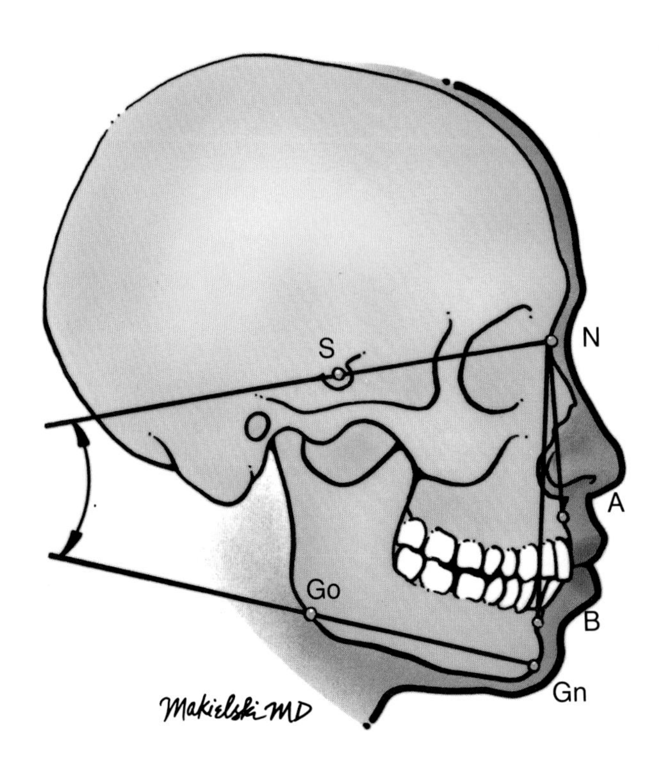

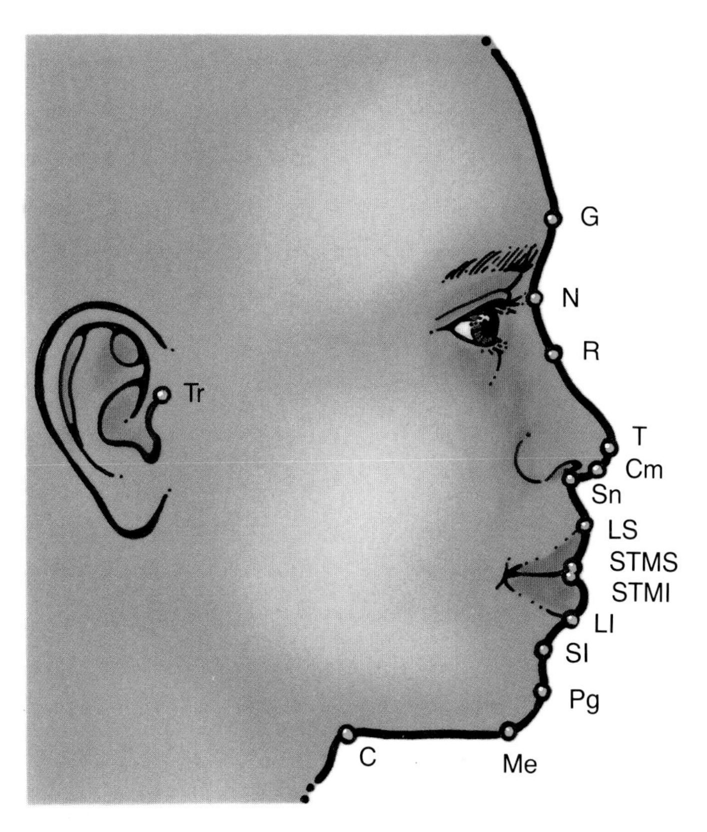

FIG. 1.5. Major soft-tissue cephalometric points. The glabella (**G**) is the most prominent point in the midsagittal plane of the forehead. The nasion (**N**) is the deepest depression at the root of the nose in the midsagittal plane. The rhinion (**R**) represents the junction of the bony and cartilaginous dorsum and is usually the maximal hump on the nose. The tip (**T**) is the most anterior projection of the nose. The columella point (**Cm**) is the most anterior soft-tissue point on the columella. The subnasale (**Sn**) is the junction of the columella with the upper cutaneous lip. The labrale superius (**LS**) represents the mucocutaneous junction of the upper lip at the midsagittal plane. Similarly, the stomion superioris (**STMS**) represents the lower border of the upper lip at the midsagittal plane. The stomion inferioris (**STMI**) and the labrale inferius (**LI**) are similarly described for the lower lip. The sulcus inferioris (**SI**) represents the deepest depression in the concavity between the lip and the chin. The pogonion (**Pg**) is the most anterior point on the chin. The menton (**Me**) is the lowest point on the contour of the soft-tissue chin. The cervical point (**C**) represents the junction between the submental area and the neck. The tragion (**Tr**) is the point at the superior aspect of the tragus.

FIG. 1.4. Steiner analysis. Go–Gn defines the mandibular plane. The angle between the mandibular plane **Go–Gn** and the **S–N** describes the vertical height of the face. **SNA** describes the anterior-posterior relationship between the maxilla and the cranial base. **SNB** similarly describes the relationship of the mandible and the cranial base. The following are typical values for Caucasians. **SNA** is usually 82 ± 2°; **SNB** is normally 78 ± 2°; **ANB** is usually 2 ± 1.5°; the mandibular plane and the **SN** create an angle of about 32°.

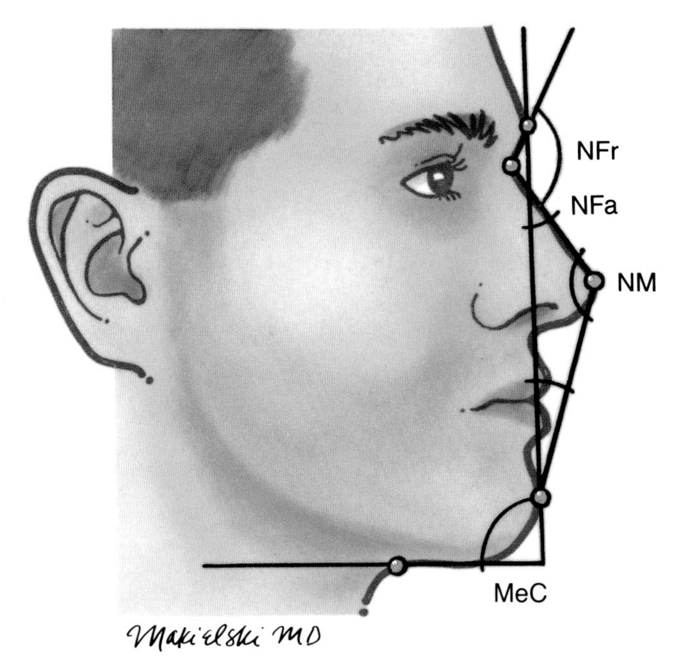

FIG. 1.6. Aesthetic triangle of Powell and Humphreys. This simple analysis is based on angles between external landmarks and, thus, can be extracted from any lateral representation (e.g., a photograph). A vertical line dropped from the glabella to the pogonion defines the vertical anterior facial plane. A line from the nasal tip is then drawn to the nasion. A line from the glabella to the nasion meets this nasal line and creates the nasofrontal angle (**NFr**). The nasofacial angle (**NFa**) is between the anterior facial plane and the line tangent to the dorsum of the nose. The nasomental line is drawn from the nasal tip to the pogonion and creates the nasomental (**NM**) angle. A line drawn from the cervical point to the menton intersects the anterior facial plane and creates the mentocervical angle (**MeC**). In Caucasians, typical "ideal" ranges are **NFa** = 30–40°, **NM** = 120–130°, **NFr** = 115–130°, **MeC** = 80–95°. Standard values for other races have not been defined. Males have more prominent features (larger **NFa,** smaller **NM, NFr,** and **MeC**) than do females.

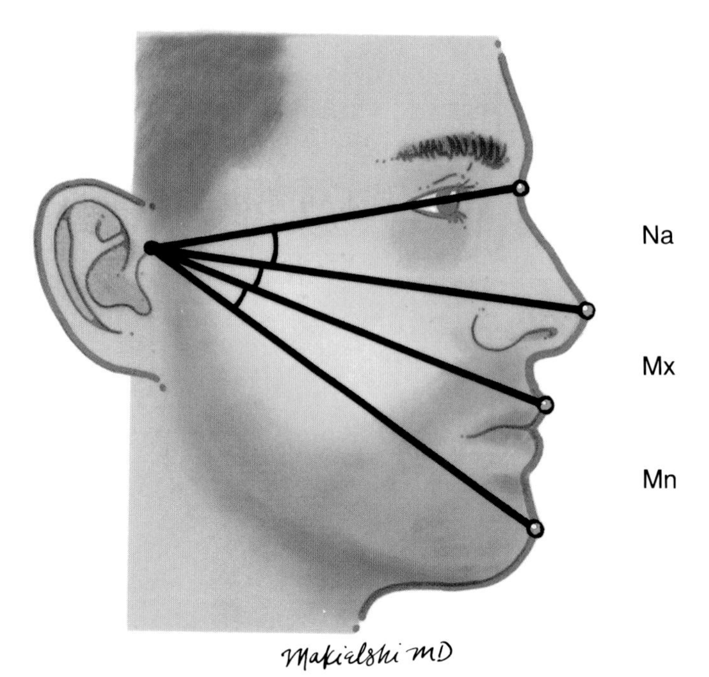

FIG. 1.7. Nasal, maxillary, and mandibular angles of Peck and Peck. Peck and Peck describe a nasal angle (**Na**) that measures the nasal height from the nasion to the tip, a maxillary angle (**Mx**) that measures the maxillary height from the tip to the labrale superius, and a mandibular angle (**Mn**) that records the mandibular height from the labrale superius to the pogonion. In their study, the mean values for these angles in adults were 23.3°, 14.1°, and 17.1°, respectively.

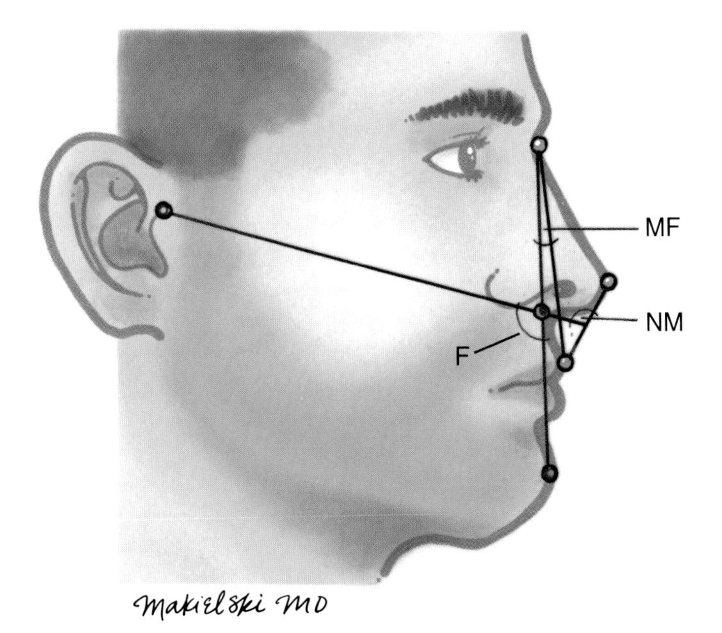

FIG. 1.8. Facial, maxillofacial, and nasomaxillary angles of Peck and Peck. Peck and Peck describe a unique orientation plane. A line from the nasion to the pogonion is bisected by a line extended from the tragion to create the orientation plane. The point where these lines cross describes a facial angle (**F**). The maxillofacial angle (**MF**) is determined by extending another line from the nasion to the labrale superius. This angle relates the upper lip to the chin. A final line from the labrale superius to the nasal tip creates an angle with the orientation plane termed the nasal maxillary angle (**NM**), which relates the upper lip to the nasal tip. For Caucasians, mean values are **F** = 102.5°, **MF** = 5.9°, and **NM** = 106.1°.

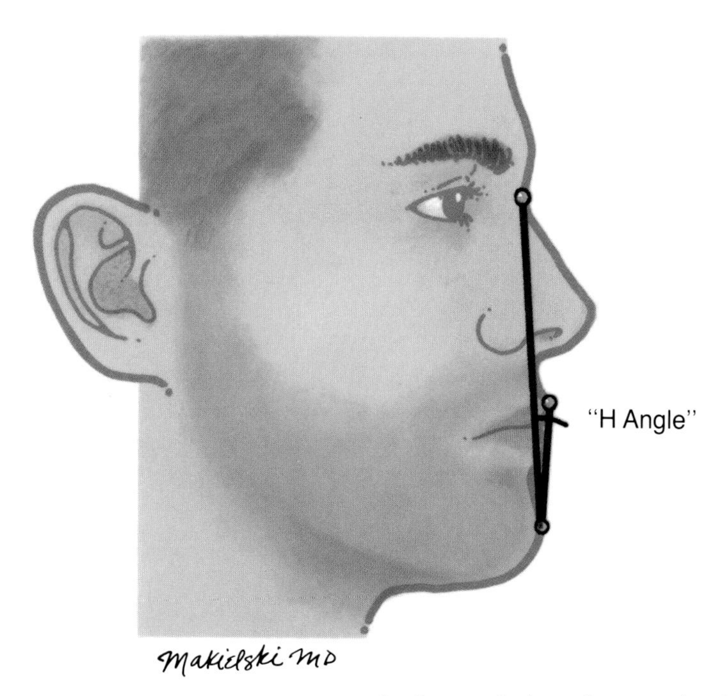

FIG. 1.9. The H angle of Holdaway. The "harmony line" extends from the pogonion to the most prominent part of the upper lip. The soft-tissue facial line runs from the soft-tissue nasion to the pogonion to create the **H** angle. A normal **H** angle is 10° and a larger **H** angle relates to increasing soft-tissue convexity of the face. The **H** angle represents a single soft-tissue measurement of maxillary-mandibular relationships.

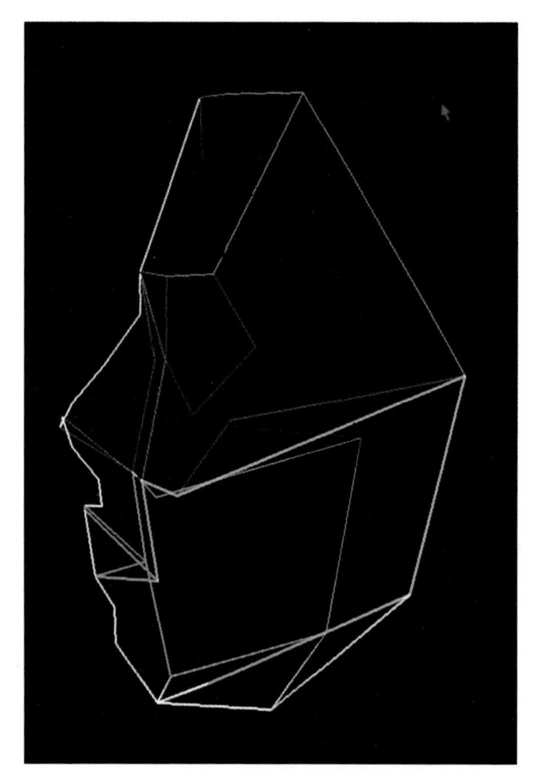

FIG. 1.10. Three-dimensional facial analysis. Standard soft-tissue cephalometric systems are designed primarily around two-dimensional analysis of the facial profile. Although useful, particularly in evaluating maxillary-mandibular relationships, these systems do not allow analysis of complex contours such as those of the malar eminence. The above graphic was generated by interconnecting key cephalometric points input to a microcomputer with a three-dimensional pointing device. This allows both visual examination of a complex data set and analysis of any number of three-dimensional relationships. (From Larrabee et al., 1988, with permission.)

CHAPTER 2

Variations in Facial Anatomy with Race, Sex, and Age

VARIATIONS WITH RACE

It is important to recognize racially and ethnically appropriate features when performing facial surgery. Although certain generalizations about race and anatomy are possible, individual variation is large. Most of the cephalometric systems have been developed in Western Europe and the United States, and therefore their average values for measurements reflect a Caucasian population. In the absence of a generally accepted racial classification, we will, for consistency and simplicity, refer to individuals of African descent as African, individuals of Asian descent as Asian, and individuals of European descent as Caucasian. The Native Americans represent a diverse group of peoples; their facial anatomy is generally closest to that of the Asian group.

The most well-studied of these racial differences are those in nasal anatomy. Hinderer (1971) described three nasal types with typical racial characteristics: platyrrhine (African), mesorrhine (Asian), and leptorrhine (Caucasian) (Fig. 2.1). He used the nasal index (a ratio of the nasal width between the piriform crests and the length of the nose) and the tip index (the ratio of the width of the nose at the nostril apex to the width at the widest expansion of the ala) to measure these differences; these indices had been used previously in anthropological literature. These observable differences are based on specific anatomic differences; for instance, the African nose has thicker skin, a smaller nasal spine, and less cartilage support than the Caucasian nose.

The mandibular-maxillary and dental relationships are different among the races and have been documented cross-culturally with cephalometric studies. A cephalometric study of Caucasian children, "the Alabama analysis," has been the basis for several comparative studies of non-Caucasian populations (Taylor and Hitchcock, 1966; Alexander and Hitchcock, 1978). Both Chinese children (Guo, 1971) and Japanese children (Miura et al., 1965) have a retroposition of the mandible and more labially inclined incisors than Caucasians. African-American children have a somewhat more anteriorly placed maxilla and more protrusive incisors than Caucasians (Drummond, 1968). Mexican-American children have a more protrusive mandible and minor differences in the incisors (Garcia, 1975). None of these studies showed any significant sexual differences within the same populations. Again, it should be emphasized that individual variation is far more significant than the differences in population averages.

Other facial features show additional racial variations. The external ear of Africans is shorter than that of Caucasians, whereas the Asian ear is generally longer. African lips are fuller than either Asian or Caucasian; Asian lips are often somewhat fuller than Caucasian.

VARIATIONS WITH SEX

There are clear anatomic differences between men and women in specific facial areas as well as different standards of beauty. One obvious difference is in hair distribution, which becomes important in designing flaps such as forehead-lift and face-lift flaps. For these cases, an evaluation of the hair density, the shape of the hairline, and (in men) the beard density and distribution is necessary. The female brow tends to be more arched than the male; the highest point is normally between the lateral limbus and the lateral canthus. The male brow is more horizontal.

Other obvious differences between men and women appear in profile analysis. In the neck, the thyroid cartilage in men is more prominent than in women. When asked to judge aesthetic profiles, observers tend to prefer men to have a more prominent nose and chin and a more acute nasolabial angle than comparative women (Lines et al., 1978).

CHANGES WITH AGING

The sequence of changes that occur in the aging face is relatively uniform; however, the rate of change varies from person to person. About age 30, sagging of the facial skin first becomes apparent, particularly where the upper eyelids overlap the palpebral lines. Also, the mesolabial folds deepen. At about 40 years of age, forehead wrinkles and horizontal skin lines at the lateral canthus begin to appear, and undulation of the mandibular line becomes noticeable. At age 50, the lateral canthus begins to slant downward, the nasal tip starts to descend, and wrinkles appear in the perioral area and the neck. About the same time, some resorption of adipose tissue in the temporal and the cheek areas occurs. At 60 years of age, the illusion of decreased eye size becomes pronounced, the skin is thinner, and fat resorption in the buccal and the temporal areas is more marked. By 70 years of age and thereafter, all these changes combine with progressive resorption of subcutaneous fat (Gonzalez-Ulloa et al., 1971). These changes can be clearly seen in the sequential self-portraits of Rembrandt (Fig. 2.2).

Regional Changes with Age

Other changes take place in specific areas of the face. The skull becomes thinner and smaller with age, causing an excess of overlying facial tissue. Beginning at age 25, the eyebrows steadily descend from a position well above the supraorbital rim to a point far below it; sagging of the lateral aspect of the eyebrows makes the eyes seem smaller. The excess of skin above the eyes combined with a weakening of the orbital septum allows intraorbital fat to herniate and create palpebral bags. Progressive descent of the nasal tip with age causes the upper and the lower lateral cartilage to separate, thus enlarging and lengthening the nose. Resorption of alveolar bone results in a relative excess of soft tissue in the perioral area.

The chin descends in much the same manner as the nasal tip and the brows. The well-defined angle between the submandibular line and the neck is lost with age. The hyoid bone and the larynx gradually descend, making the larynx look more prominent. The cervical appearance with aging is a combination of changes in the skin, fat distribution, the platysma muscle, and the underlying bony/cartilaginous framework. The anterior edges of the platysma separate and lose tone; this creates the anterior banding so characteristic of the aging neck. Fat is frequently deposited in the submental area. This fat, combined with laxity of the skin, causes a loss of the cervical-mental angle (Fig. 2.3).

Changes in the Skin

Most of the changes in the external appearance of the face are the result of gravity acting on skin that is becoming progressively thinner, drier, and less elastic. The skin itself shows increased wrinkling and pigmentary changes with age. Overexposure to sunlight hastens the skin changes and thus speeds up the aging process. Genetic factors influence the localization and the shape of facial wrinkles and the age at which the hair turns gray and alopecia develops.

Electron microscopic examination of the undersurface of aged skin shows a general loss of epidermal complexity (Fig. 2.4). The epidermis of wrinkles is flattened and has few microvilli, whereas the basal cells surrounding wrinkles contain the normal dense microvilli (Larrabee et al., 1984). The delicate elastic fiber network characteristic of young skin becomes dense and less organized with the passage of time. Collagen synthesis and degradation both decrease with age; in addition, the collagen in aged skin is probably more stable (increased number of nonreversible cross-links) than that in younger skin.

Many of the gross and microscopic changes seen in aged skin also occur in young skin exposed excessively to sunlight. With the electron microscope one sees a flattening of the undersurface of the skin and degeneration of the architecture of the fine elastic fibers and the blood vessels in aging skin (Montagna and Carlisle, 1979). Skin from various sites on the face is thinner in aged persons than in young adults (Gonzalez-Ulloa and Flores, 1965).

As water-binding capacity and sebaceous gland activity decrease with age, the skin becomes drier. Sebaceous gland activity is primarily related to androgen production; sebum production falls steadily in women after menopause but remains fairly stable in men until about age 70 (Gilchrist et al., 1983).

These histologic and biochemical changes translate into an aging skin that is less elastic and has a more irregular and pigmented surface than young skin.

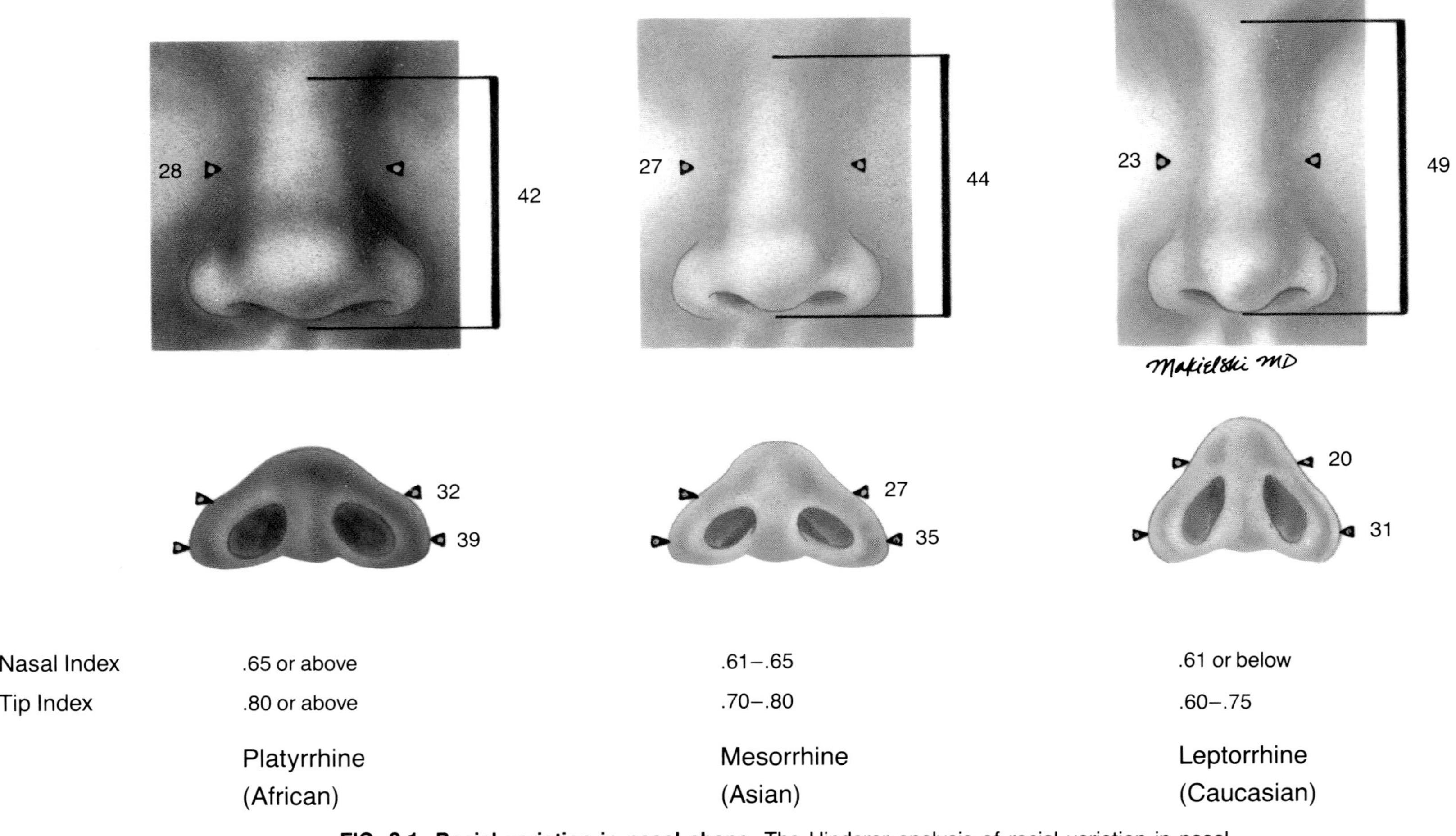

FIG. 2.1. Racial variation in nasal shape. The Hinderer analysis of racial variation in nasal shape, drawn from the anthropological literature, uses the "nasal index" (a ratio of the nasal width between the piriform crests and the length of the nose) and the "tip index" (the ratio of the width of the nose at the nostril apex and the width at the widest part of the ala). The African nose is both wider and shorter than the Caucasian, resulting in a larger nasal index; the Asian nose is intermediate. The African nose has a less triangular shaped nasal base and thus a higher tip index than the Caucasian nose; again the Asian nose has an intermediate shape. These variations in shape are based on the underlying anatomic differences in nasal structure among the races. Caucasians have, on average, thinner skin, more prominent tip cartilages, a more prominent nasal spine, and longer nasal bones than Africans. As in the external appearance, the Asian nose is intermediate in structure.

FIG. 2.2. Self-portraits of Rembrandt van Rijn demonstrating progressive changes with age. A: 1628–29: Rembrandt as a young man. (© Indianapolis Museum of Art, The Clowes Fund Collection.) **B:** 1636–38: The upper eyelids overlap the lid fold, the melolabial lines deepen, forehead lines appear, and undulation of the mandibular line becomes noticeable. (© The Norton Simon Foundation, detail of Rembrandt.) **C:** 1660: The brows and the nasal tip descend, and the rhytids of the forehead, the perioral area, and the neck deepen. (© The Metropolitan Museum of Art, Bequest of Benjamin Altman, 1913, detail of Rembrandt.) **D:** 1669: The jowls increase and there is loss of cervicomental angle as well as progression of the previous changes. (© Mauritshuis–The Hague.)

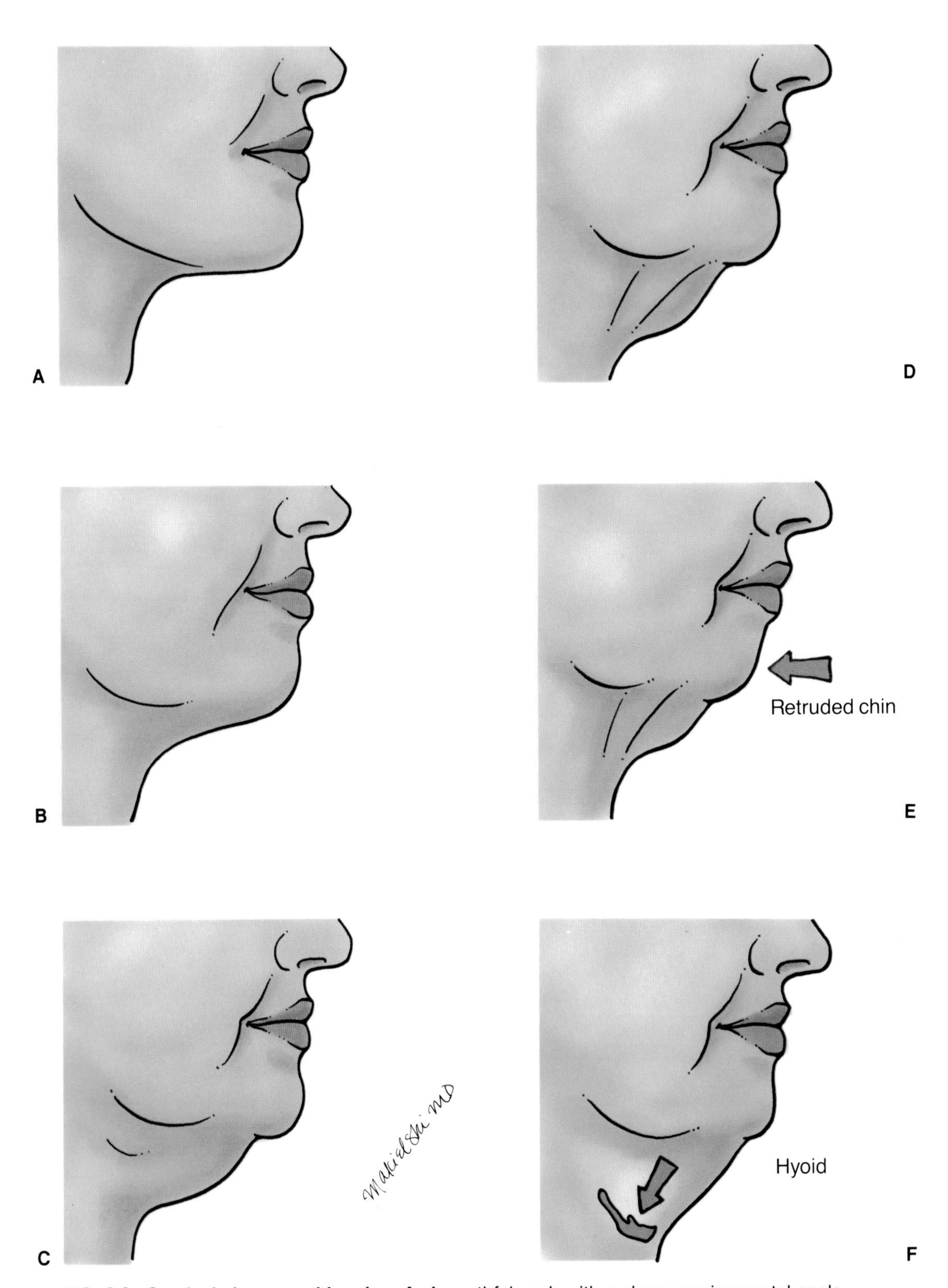

FIG. 2.3. Cervical changes with aging. A: A youthful neck with a sharp cervicomental angle and a strong mandibular line. **B:** Early loss of the cervicomental angle and the mandibular line. **C:** Chin ptosis, accumulation of submental and submandibular fat, laxity of submental skin. **D:** Accentuation of the previous changes plus banding of the anterior platysma. **E:** Further descent and retrusion of the chin (anterior mandibular recession). **F:** Descent of the hyoid accentuates the above changes. (After Dedo, 1988.)

FIG. 2.4. Aging skin. A split skin preparation of the epidermal undersurface of an aged individual demonstrating generalized flattening and wrinkle formation. (From Larrabee, 1984, with permission.)

SECTION II

Anatomic Systems

CHAPTER 3

Hard-Tissue Foundation

The contour of the face is primarily determined by the underlying bony structure. As an outgrowth of craniofacial surgery, surgeons have become considerably more aggressive in contouring the facial skeleton for both functional and cosmetic purposes. Figure 3.1 shows the skull with key foramina.

DEVELOPMENT OF THE CRANIOFACIAL SKELETON

Like the long bones, the base of the skull develops by endochondral ossification; it first forms in cartilage, which later ossifies. In contrast, the sides and the roof of the skull and the facial bones form primarily by ossification of a membranous matrix of mesenchyme. The regenerative abilities of this membranous bone of the face and the skull are very limited (except in the young), in comparison to that of the endochondral long bones.

The ossification of the skull is incomplete at birth. The two fontanels are the last areas to ossify: the diamond-shaped anterior fontanel at the junction of the frontal, the sagittal, and the coronal sutures and the triangular posterior fontanel at the junction of the sagittal and the lambdoid sutures. These ossify during the first year of life.

A child's skull grows slowly until about age 12 years. The maxilla and the mandible are small in proportion to the skull in infancy, but increase in size and relative proportion as the child grows to adulthood (Figs. 3.2 and 3.3). The facial skeleton, including the teeth, is not fully developed until the late teen years. The primary dentition of 20 teeth is present in children from $2\frac{1}{2}$ to $5\frac{3}{4}$ years of age. The permanent dentition of 32 teeth (including the third molars) appears prior to adulthood.

ARCHITECTURE OF THE CRANIOFACIAL SKELETON

The bony vault of the cranium, or calvarium, consists of the frontal bone, the two parietal bones, and the occipital bone. Laterally the greater wings of the sphenoid lie anterior and the temporal bones posterior. The temporal bones are considerably thinner than other bones of the skull. This thinner bone is protected by the overlying temporalis muscle. Bone flaps are frequently hinged in the temporal area where they can be easily fractured. There is large variability in calvarial thickness; it averages 7 mm, with a range of 3 to 12 mm (Pensler and McCarthy, 1984).

Cranial bone consists of an outer table of compact cortical bone, a middle layer of coarse cancellous bone, and an inner table of cortical bone. The inner table is thinner than the outer table. There is periosteum on both sides of the cranial bone. The inner periosteum is fused with the dura.

The mandible is essentially a long bone bent in a "U-shaped" configuration (Fig. 3.4). It has outer and inner cortical plates that are thicker anteriorly and along its inferior border. The mandible articulates with the skull base at the synovial joints of the condyles bilaterally. Several aspects of the mandible make it a unique structure: the importance of the teeth and their occlusion, the synchronous movement of the condyles with respect to the base of the skull, and the complexity of its multiple muscle attachments.

The midface is a clinical entity that encompasses the middle third of the facial skeleton. This region can be defined as an area between the maxillary teeth and a line joining the two zygomaticofrontal suture lines. Posteriorly the limits of the midface are the sphenoethmoid junction and the pterygoid plates. Bones comprising the midface include the maxillae, the palatine bones, the zygomatic or malar bones, the nasal bones, the zygomatic processes of the temporal bones, the lacrimal bones, the ethmoid bone, and the turbinates. The alveolar process of the maxillary bone contains the maxillary teeth and is weaker than the mandible.

The teeth and their supporting structures are the key to understanding and treating abnormalities in maxillary-mandibular relationships, whether they derive from trauma or some other etiology. Use of standard nomenclature (Fig. 3.5) and understanding of the universal numbering system (Fig. 3.6) facilitate communication with dental colleagues. The simplest and most widely used classification of the anterior-posterior relationships of the teeth is that of Angle (Fig. 3.7).

BIOMECHANICS OF THE FACIAL SKELETON

The face can be analyzed as a series of arches with intervening areas of thin cortical bone. These arches provide strong protection to the vital sensory organs and the cranial contents. An arch will absorb considerable compressive force on its convex outer surface prior to fracturing; the inner cortex will likely fracture first. Thus, one may see fractures of the inner table of the frontal bone with the outer table intact. Although the mechanics of an external force acting on a single arch of bicortical bone can be fairly easily understood, the assembly of these arches of various shapes, thicknesses, and positions that comprise the human face defies a simple description. The horizontal and the vertical facial buttresses represent the best available understanding of the mechanical supports of the face (Figs. 3.8 and 3.9). These buttresses determine how an impact force is distributed over the face. Although we have included the standard central vertical buttress, the force vectors may actually skirt the frontal sinus and follow the orbital rims.

COMMON SITES OF FACIAL FRACTURE

Fractures of the facial bones occur at predictable sites based on the anatomic location and the vector of force. After a fracture, the degree and the direction of bony displacement is determined to a large degree by the pull of the associated muscles. The relative incidence of mandibular fractures in specific regions is depicted in Figure 3.10. The influence of the muscles on fracture displacement is shown in Figures 6.16 and 6.17. Midface fractures are frequently described using the Lefort classification, although the variability of exact fracture sites and differences from side to side argue for a simple description of these fractures rather than attempting to make them fit a Lefort I, II, or III pattern. The Lefort classification is historically significant, however, and provides a useful general classification for midfacial fractures. Lefort I is a fracture of the maxillary dentoalveolar complex; the teeth and the alveolar process are mobile, but not the remainder of the midface. Lefort II is a pyramidal fracture in which the middle third of the face is mobile, but the malar complex is stable; Lefort III represents a craniofacial disjunction in which the entire face is mobile with respect to the skull.

Nasal fractures are the most common facial fracture and usually occur over the thinner aspects of the nasal bones (Fig. 14.22). With great force the nasal bones can be comminuted and telescoped into the fragile ethmoid complex to create a nasoethmoid complex fracture.

The globe is well protected by the thick orbital rims. Direct blows to the eye will not usually cause a fracture of the orbital rim, but will more commonly result in a "blow-out" fracture of the thin orbital floor. Because of the close association of the infraorbital nerve and the inferior rectus muscle with the orbital floor, the patient with a blowout fracture may experience a loss of sensation over the cheek and muscle entrapment, which limits upward gaze.

The most common site of fracture in the lateral midface is the malar bone. Malar fractures vary from a simple depression of the zygomatic arch to the classic "tripod" fracture, which involves fracture sites at the zygomaticofrontal suture, the infraorbital rim (usually the weak infraorbital foramen), and the arch. The trimalar fracture tends to displace inferiorly and medially because of the masseter muscle attachment (Fig. 6.12). Malar fractures, like Lefort fractures, can be quite variable and do not always follow the classic pattern.

AUGMENTATION AND GRAFTING OF THE FACIAL SKELETON

The face is unique in that essentially the entire bony framework can be augmented or reduced with relatively inconspicuous scarring through a combination of the coronal incision and various intraoral incisions (Figs. 3.11 and 3.12). Augmentation or grafting may be required in circumstances such as reconstruction after trauma or cancer surgery. Autogenous bone provides the best current grafting material. Graft survival and osteogenesis are better with cancellous bone than with cortical bone due to the presence of more osteoprogenitor cells. Many sites have been used in the past to harvest bone grafts, particularly the rib and the iliac crest. Currently, the outer table of the calvarium is the preferred site in most cases. These cranial bone grafts can be harvested in the same field as the surgery and leave a minimal defect (Fig. 3.13). Cranial bone is membranous bone and undergoes less resorption than endochondral bone (Smith and Abramson, 1974; Zins and Whitaker, 1983).

Autogenous bone grafts survive by osteoconduction. They are first vascularized and then demineralized, and then bone growth occurs by conduction from the surrounding bone. In contrast, osteoinduction transforms undifferentiated mesenchymal cells into osteoblasts to create new bone. Recent studies have demonstrated that osteoinductive bone factors can create clinically useful bone for craniofacial reconstruction. The two most likely candidates at this time are bone morphogenic protein (BMP) and transforming growth factor beta (TGF-B) (Toriumi et al., 1991). These recombinant osteoinductive compounds hold the promise of dramatically improving our ability to augment and to strengthen the craniofacial skeleton.

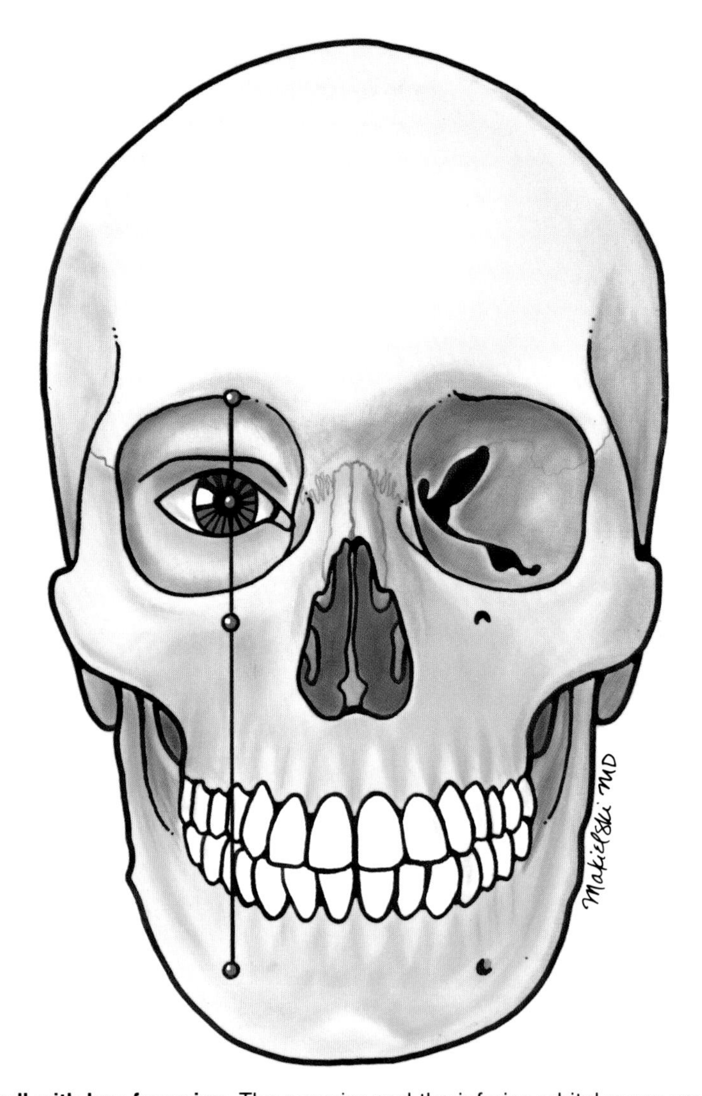

FIG. 3.1. Skull with key foramina. The superior and the inferior orbital nerve are seen on the left. The supraorbital nerve (V1), the infraorbital nerve (V2), and the mental nerve (V3) represent the major cutaneous nerves of their divisions and emerge from the foramen approximately in a sagittal plane at the midpupillary line. The supraorbital nerve can emerge from either a notch or a foramen, which is frequently palpable. It lies approximately one fingerbreadth lateral to the nose. The infraorbital foramen usually lies just slightly lateral to the supraorbital foramen. The mental foramen usually falls in a vertical line from the lateral edge of the supraorbital foramen or the notch.

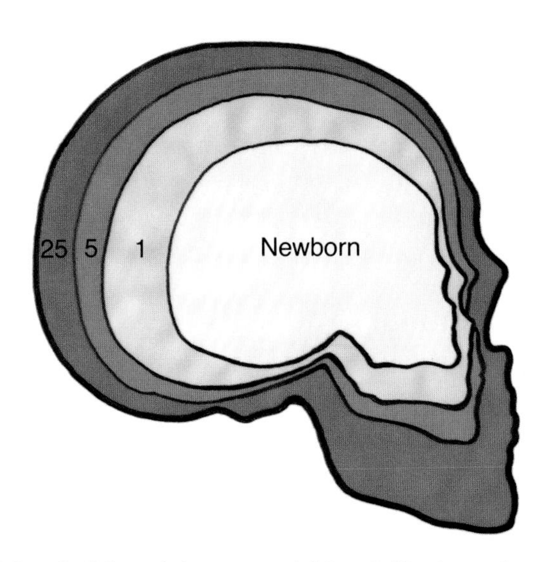

FIG. 3.2. Growth of the skull from infancy to adulthood. Newborn, 1 year, 5 years, 25 years. The majority of the skull's growth takes place in the first 5 years of life with only slow changes from 5 years to maturity. Skull growth is essentially complete by age 12. (After Lowrey, 1986.)

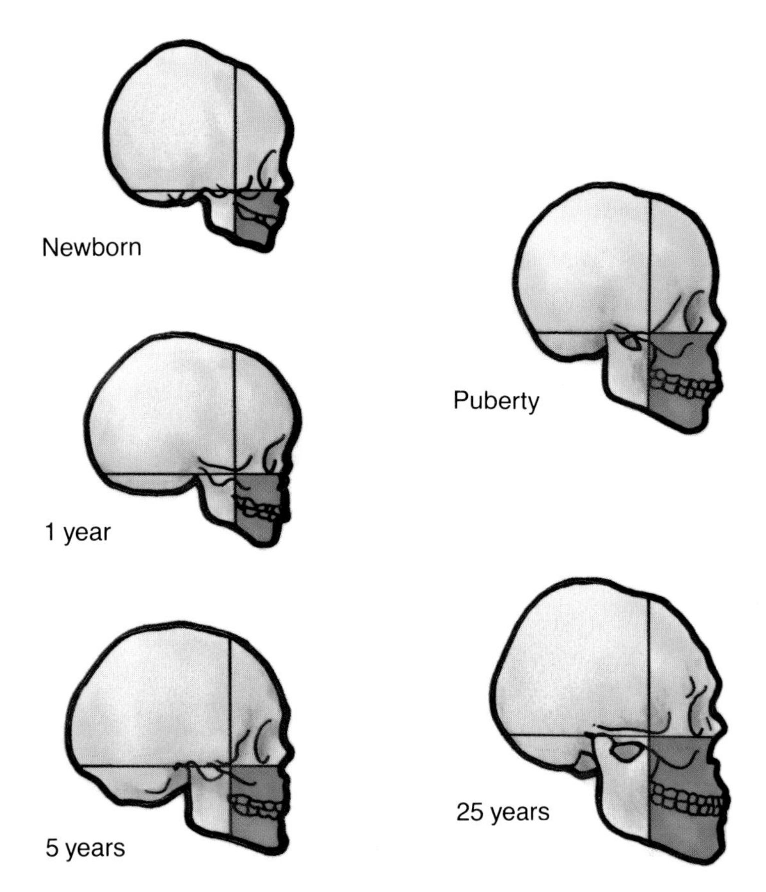

FIG. 3.3. Growth of the face relative to the skull from infancy to adulthood. Newborn, 1 year, 5 years, puberty, 25 years. At birth, the skull is relatively large and mature when compared to the face. The relative proportions between the face and the skull continue to change until adulthood as the face becomes more and more prominent. Final dentition (the third molars) is attained prior to adulthood. (After Lowrey, 1986.)

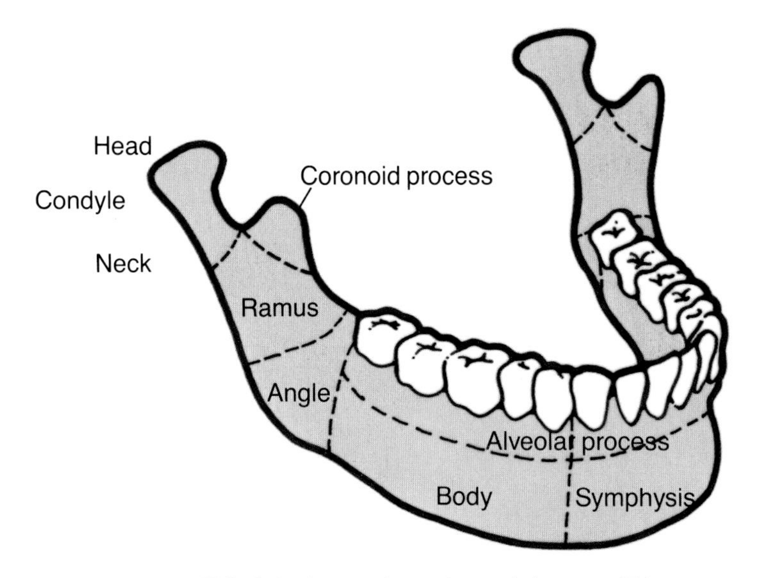

FIG. 3.4. Anatomic regions of the mandible.

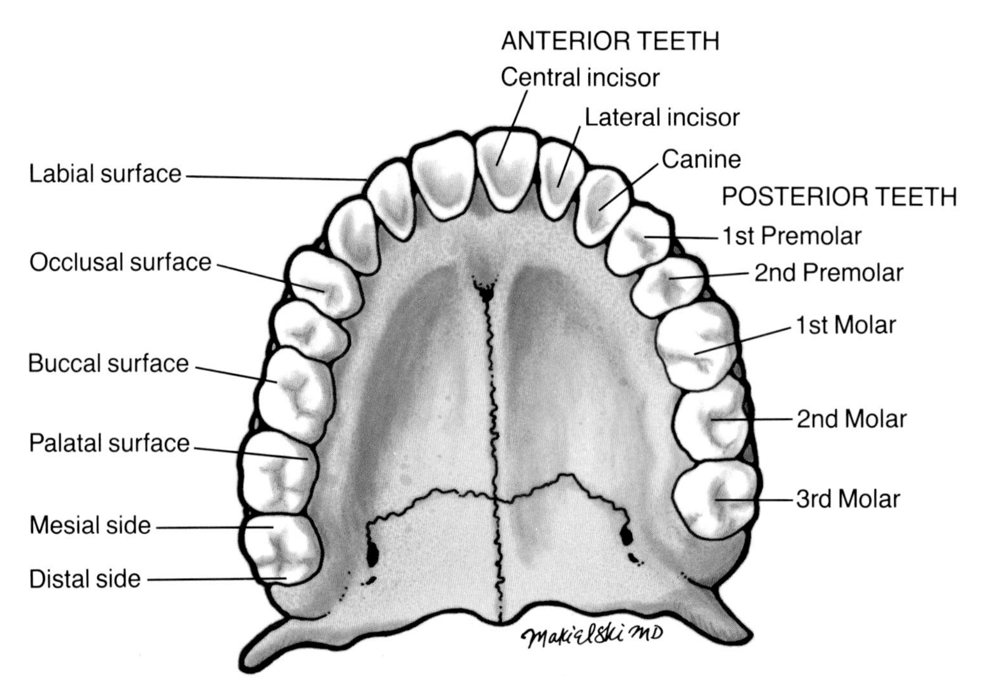

FIG. 3.5. Dental nomenclature. The major descriptive terms for the teeth are included. The entire inner surface of the teeth is referred to as the palatal surface in the maxilla and as the lingual surface in the mandible.

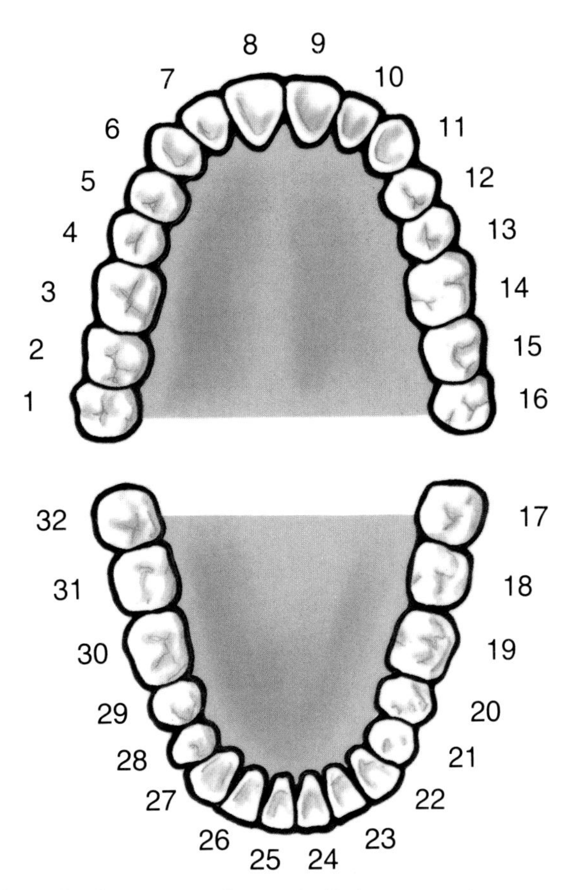

FIG. 3.6. Universal numbering system for teeth. Reference to these standard numbers facilitates communication between professionals involved in a patient's care.

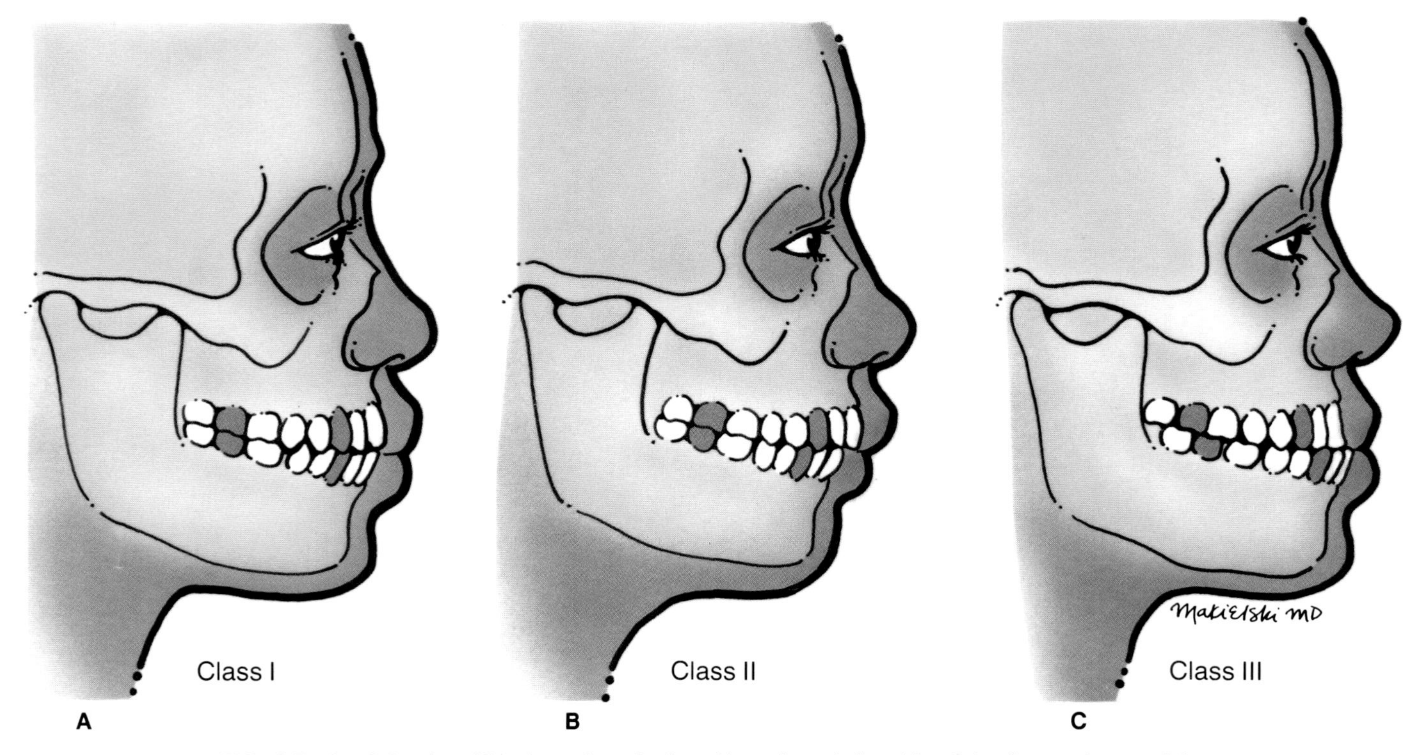

FIG. 3.7. Angle's classification of occlusion. Note the relationship of the first molars and the canines (in blue). **A: Class I—orthognathic.** The mesiobuccal cusp of the maxillary first molar rests in the mesiobuccal groove of the mandibular first molar, and the maxillary canine occludes with the distal half of the mandibular canine and the mesial half of the mandibular first bicuspid. **B: Class II—retrognathic.** The buccal groove of the mandibular first molar is distal to the mesiobuccal cusp of the maxillary first molar, and the distal surface of the mandibular canine is distal to the mesial surface of the maxillary canine. **C: Class III—prognathic.** The buccal groove of the mandibular first molar is medial to the mesiobuccal cusp of the maxillary first molar, and the distal surface of the mandibular canine is mesial to the mesial surface of the maxillary canine.

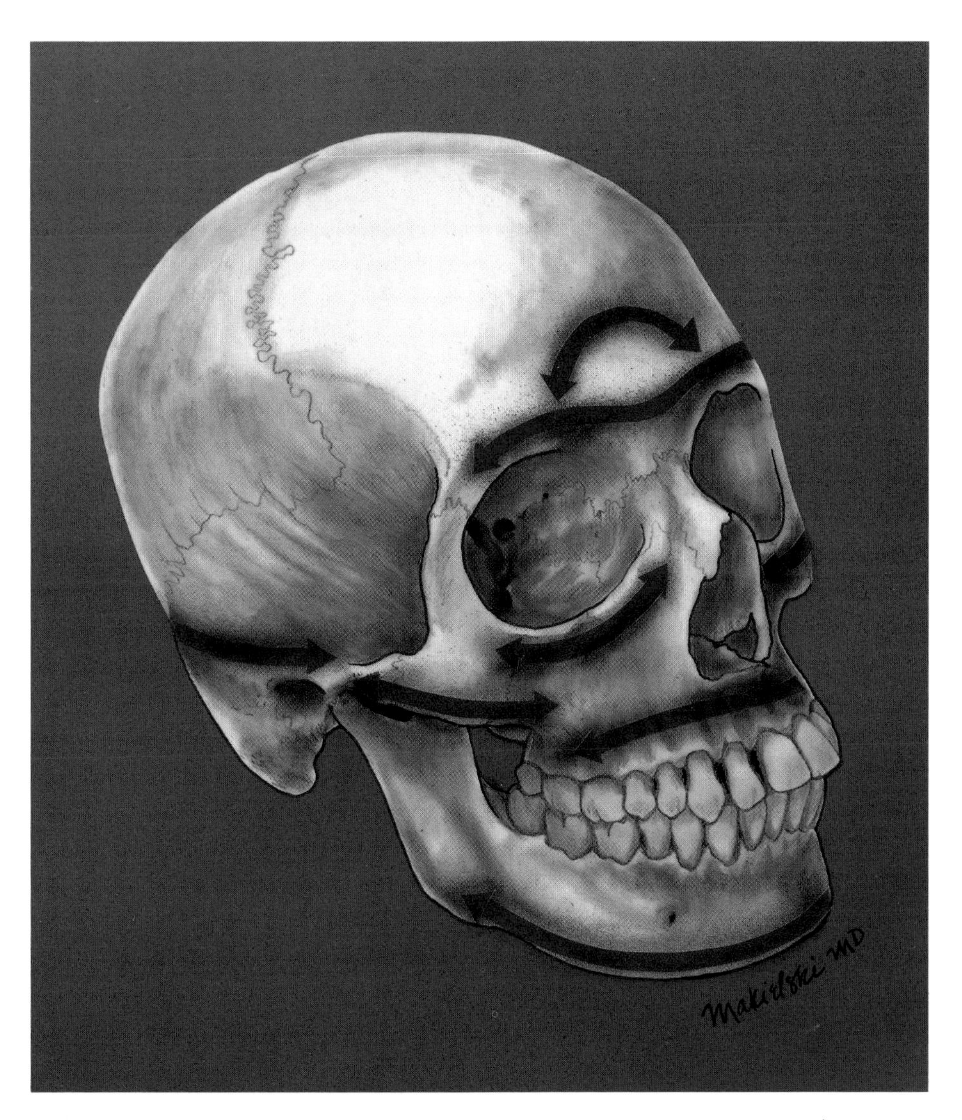

FIG. 3.8. Horizontal buttresses of the skull. The purple areas represent areas of thicker facial bone that are less likely to fracture than intervening areas.

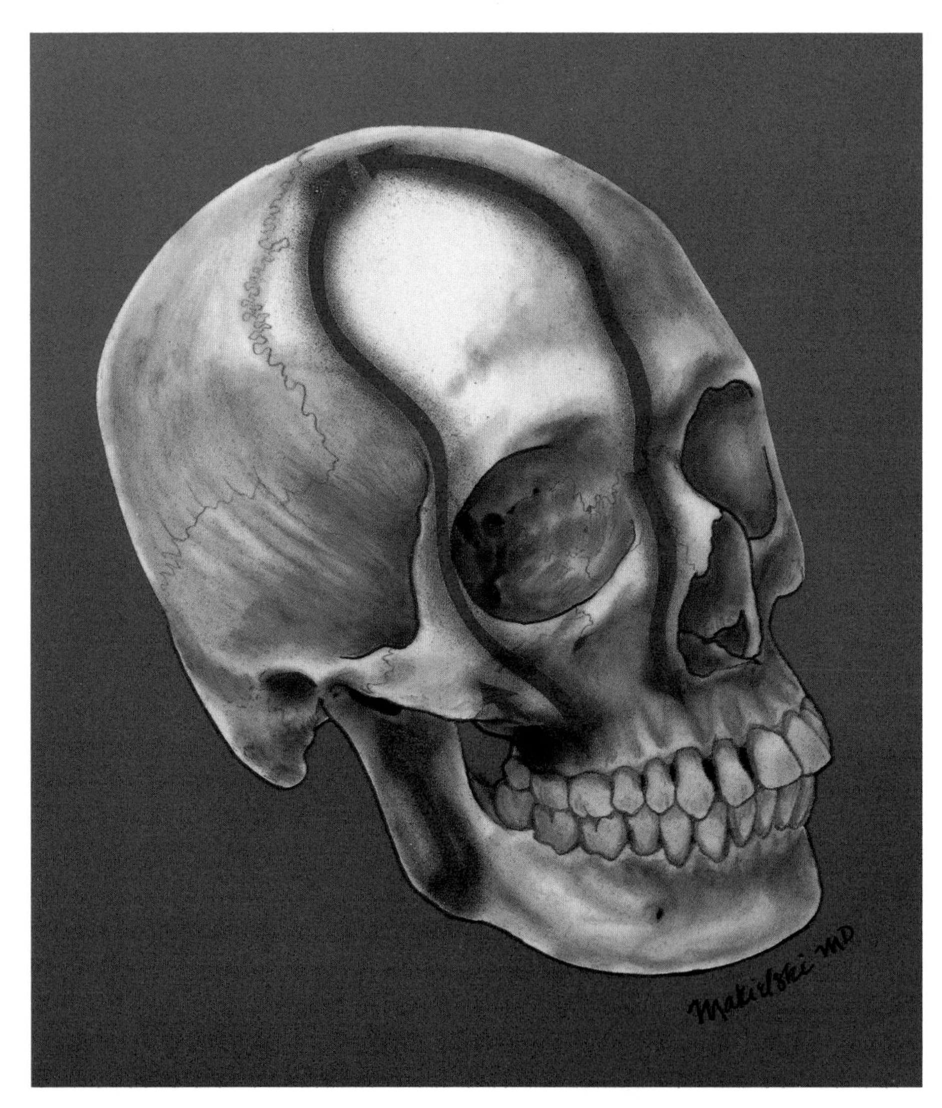

FIG. 3.9. Vertical buttresses of the skull. The purple areas represent areas of thicker facial bone that are less likely to fracture than intervening areas. Depending on the development of the sinuses, the buttress may follow the supraorbital rim and skirt the frontal sinus.

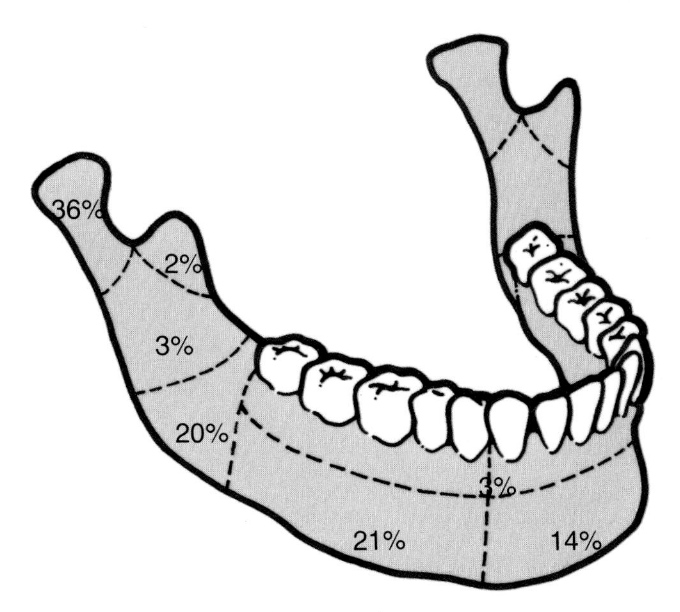

FIG. 3.10. Fracture sites of the mandible. The approximate frequencies of fracture at the indicated regions are shown.

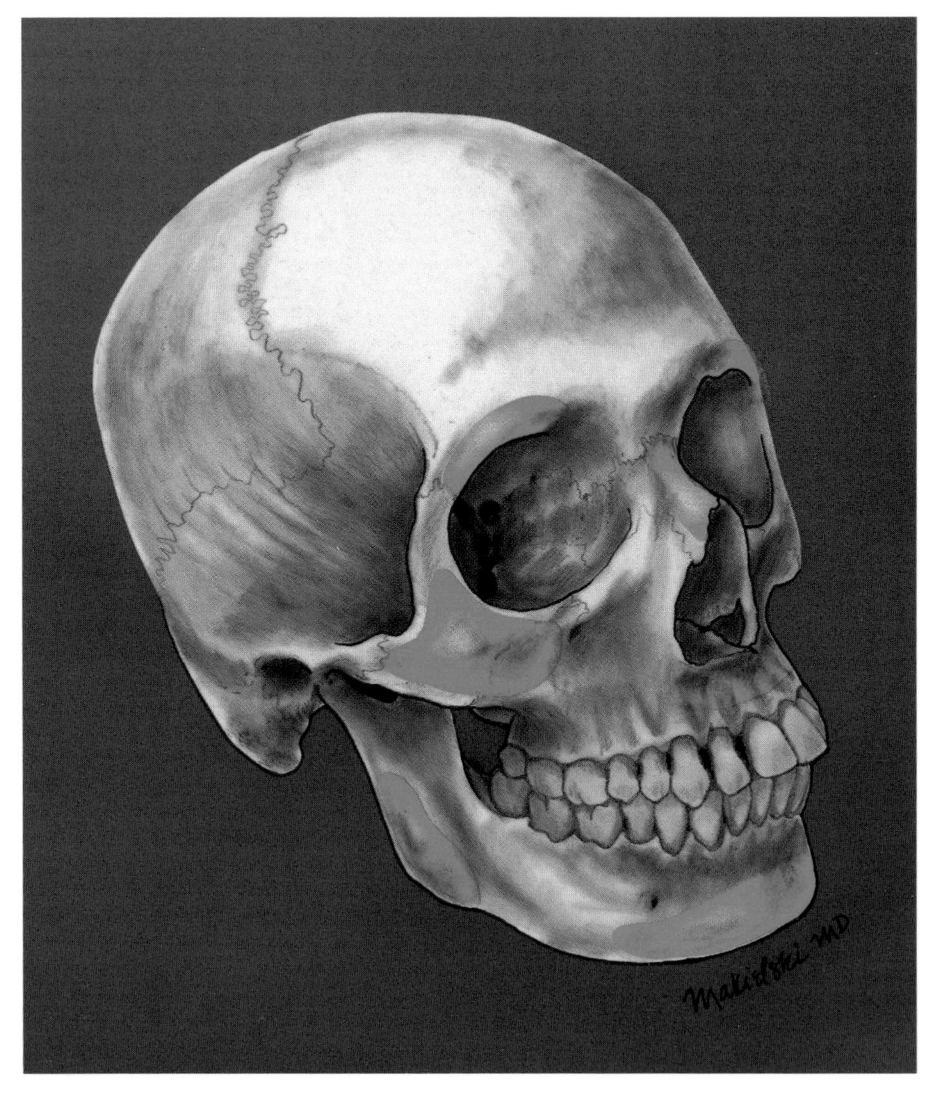

FIG. 3.11. Skull: common areas to augment or reduce. Through coronal or intraoral incisions the facial skeleton can be augmented or reduced at these key contour sites.

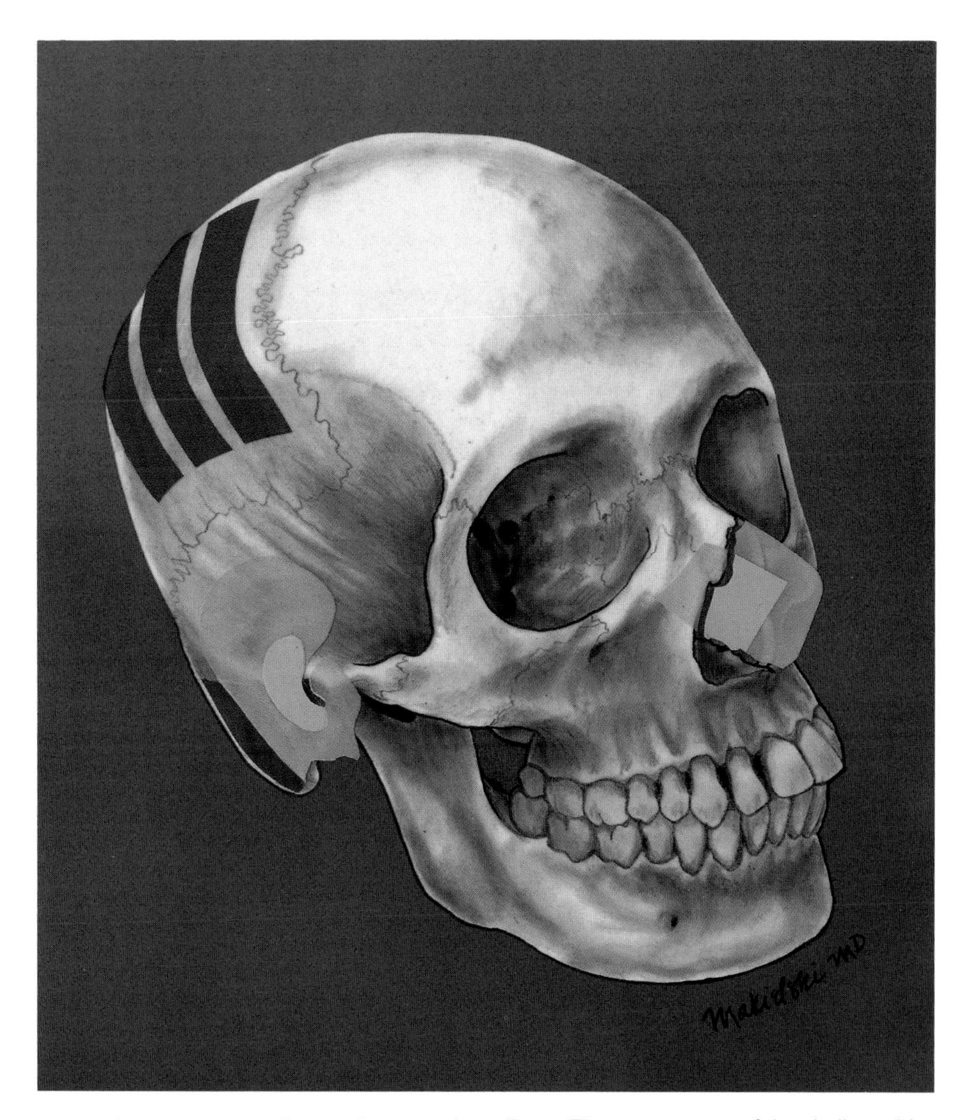

FIG. 3.12. Skull: sources of spare bone and cartilage. The outer cortex of the skull provides a readily available source of bone for craniofacial grafting. This cranial bone is of membranous origin and thus undergoes less resorption than the endochondral bone. The mastoid cortex is an excellent site to harvest smaller bone grafts, particularly for those surgeons experienced in mastoid surgery. The nasal septum and the auricle provide a convenient source for cartilage grafts in the head and the neck. The mid-portion of the septum can be safely harvested for nasal or eyelid reconstruction if 1 to 2 cm of cartilage are left in the dorsal and caudal septum to preserve support. The thicker cartilage along the maxillary crest and at the junction with the bony septum is especially useful for strong struts and tip grafts in rhinoplasty. Auricular cartilage is not as easily crafted for precise grafting as septal cartilage. The entire cymba concha and cavum concha can be harvested without causing external deformity as long as the radix helicus is preserved. The cartilage of the cavum concha toward the ear canal is thickest and is preferred for the creation of strong nasal tip grafts when septal cartilage is unavailable.

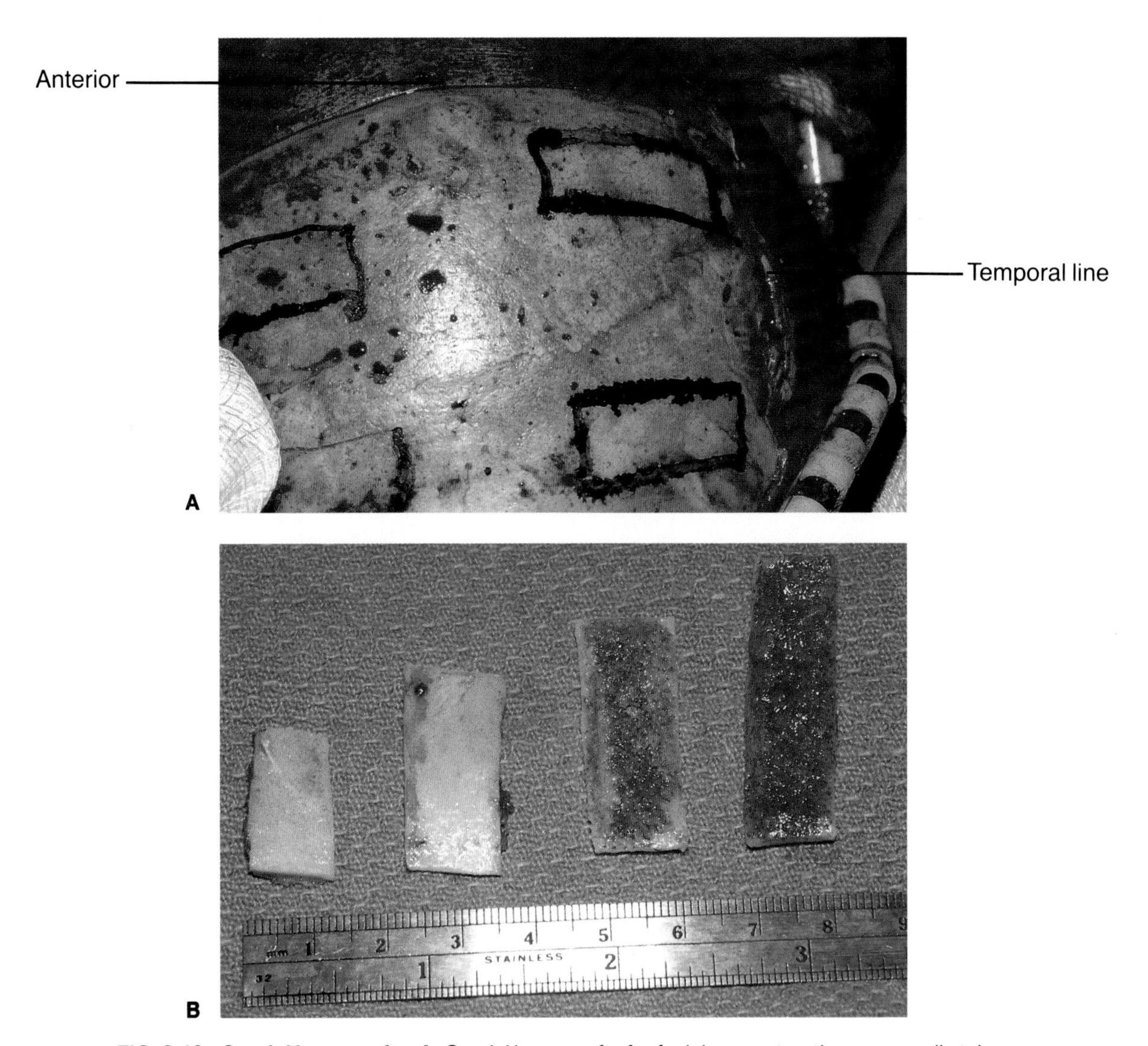

FIG. 3.13. Cranial bone grafts. A: Cranial bone grafts for facial reconstruction are usually taken above the temporal line, away from the suture lines, and at least 1.5 cm from the midline to avoid the sigmoid sinus. **B:** Harvested outer table grafts are ideal for facial reconstruction.

CHAPTER 4

Skin and Soft Tissue

The skin and the immediately underlying subcutaneous fat form a complex organ that provides soft-tissue coverage for the hard-tissue foundation of the face.

FACIAL AESTHETIC UNITS

The surface of the face can be divided up into facial aesthetic units (Gonzalez-Ulloa et al., 1954). Within each unit, the skin is consistent in color, texture, thickness, and mobility. This principle of facial units has been refined in recent years with the subunit principle. Subunits are topographic units with a predictable contour from person to person. By placing incisions at the boundaries of these subunits, final scars are minimized visually, since the eye is expecting a contour change. In a series of papers, Burget and Menick (1985, 1986) have defined the subunits of the nose and the lips. These are described in Chapters 14 and 17. Our concept of the aesthetic facial units is demonstrated in Figure 4.1. Although the forehead and the cheek are single units in terms of contour, they have areas of skin thickness variation within them, which are illustrated in Figure 4.1.

ARCHITECTURE OF THE SKIN

The skin is subdivided into the thinner, more superficial epidermis and the more fibrous dermis (Fig. 4.2). The epidermis is stratified squamous epithelium that varies greatly in depth, ranging from the thickness of the scalp to the delicate skin of the eyelid, which, at 0.04 mm, is the thinnest on the body. The dermis, composed primarily of connective tissue, also contains nerves, blood vessels, muscles, lymphatics, sweat glands, and pilosebaceous glands. The more superficial and thinner dermis is termed the papillary dermis. The deeper and thicker dermis is the reticular dermis. The dermis is composed primarily of collagen. The collagen fibers in the papillary dermis are thin and randomly arranged in comparison to the coarser bundles of the reticular dermis, which are arranged parallel to the surface. The difference between the collagen in the papillary dermis and that in the reticular dermis can be seen quite clearly when performing dermabrasion; as the surgeon abrades more deeply into the dermis, the coarseness of the reticular dermis becomes apparent.

The dermis contains a superficial vascular plexus and a deep vascular plexus. The superficial plexus, also known as the subepidermal or subpapillary plexus, runs in the papillary dermis just beneath the epidermis where it sends an arcade of capillary loops into each dermal papilla. This plexus is the source of the dermal bleeding seen immediately after the epidermis is removed with dermabrasion. The deep plexus, or dermal plexus, surrounds the skin appendages in the reticular dermis and is composed of larger vessels. The cutaneous vascular plexus is more extensively discussed in Chapter 9.

Subcutaneous fat varies greatly in thickness and texture among individuals and in the different regions of the face. It is thickest in the concavities of the cheeks, the temples, and the neck. Lobules of subcutaneous fat are divided by fibrous septae that contain vessels, nerves, and lymphatics. These septae represent a segment of the superficial musculoaponeurotic system (SMAS), described in Chapter 5. The vessels of these fibrous septae form the blood supply for the island pedicle flap, in which the skin around the flaps is completely divided, severing the subdermal plexus. The fibrous connections between the skin and the deeper structures are mechanically important. When advancing tissue and repairing facial defects, a moderate degree of undermining of these attachments decreases closing tensions (Larrabee and Sutton, 1981).

MECHANICAL PROPERTIES

The mechanical characteristics of the skin are determined by the major connective tissues of the dermis (the elastic fibers, the collagen fibers, and the ground substance). When a section of skin is stretched, the delicate elastic fibers are distorted first. If the skin is deformed more, the randomly oriented collagen fibers align with the direction of the force. Any further deformation will result in mechanical damage. Thus, once high closing tensions are reached, very little additional tissue can be recruited for wound closure, and other solutions, such as flaps or grafts, should be considered. Tension is an important variable in wound healing. Increased wound closure tension leads to decreased blood flow and impaired flap survival (Larrabee et al., 1984).

RELAXED SKIN TENSION LINES

Skin is anisotropic, which means skin tension and properties vary in different directions. This was probably first noted by Dupuytren in 1834 when he observed that a corpse stabbed with an awl had elliptical rather than round wounds. Many others, such as Langer and Kocher, have advanced our understanding of these directional skin lines and their surgical importance. Although of historical importance, Langer's lines do not correspond to our current understanding of the direction of tension in the skin. Borges (1973) describes the relaxed skin tension lines (RSTL), which are the lines of skin tension present when the skin is in a relaxed state. In most cases the RSTL follow the wrinkles of the face, but there are a few areas, such as the glabellar creases and temple, where strong muscle action can create creases in a different direction than the RSTL (Fig. 4.3). These exceptions, in general, are not clinically significant; for example, an incision made in a glabellar skin crease heals quite well. Experimental studies have demonstrated that it requires as much as twice the tension to close a wound made perpendicular to the RSTL, as compared to one made with them (Larrabee, 1986).

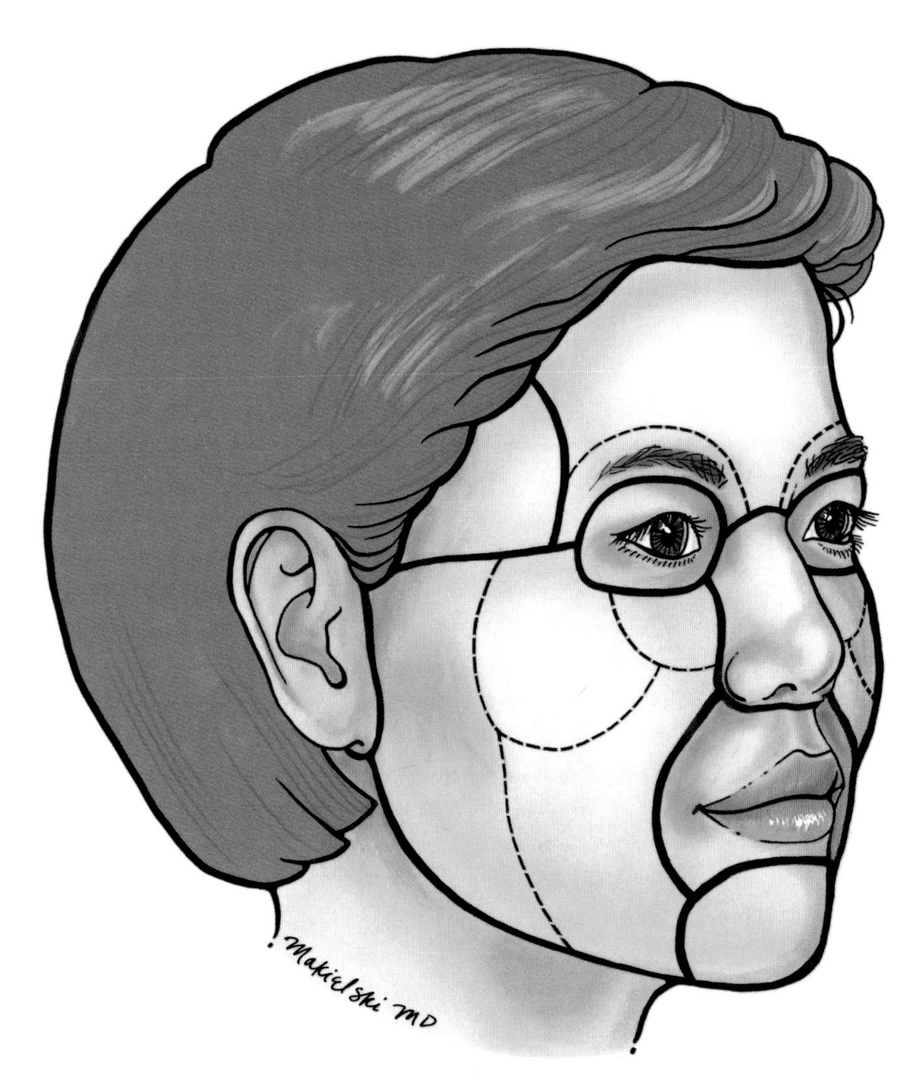

FIG. 4.1. Facial aesthetic units. The forehead, the temple, the cheeks, the nose, the periorbital area, the lips, and the chin each constitute a facial aesthetic unit that shares skin contour characteristics. Incisions that cross the unit boundaries should be avoided when possible. The forehead and the cheek have additional areas of variation in skin thickness indicated by dotted lines.

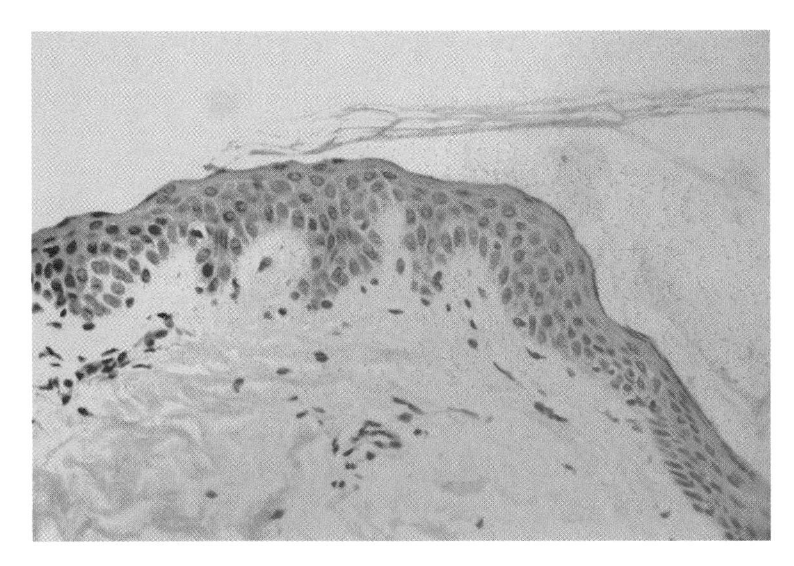

FIG. 4.2. Cross-section of the skin. The thin epidermis of stratified squamous epithelium overlies the more fibrous dermis. The collagen bundles of the superficial pupillary dermis are thinner and more randomly oriented than those of the deeper reticular dermis. During dermabrasion, the surgical depth can be determined grossly by the coarseness of the fibers exposed. (Courtesy of Susan Patterson, M.D.)

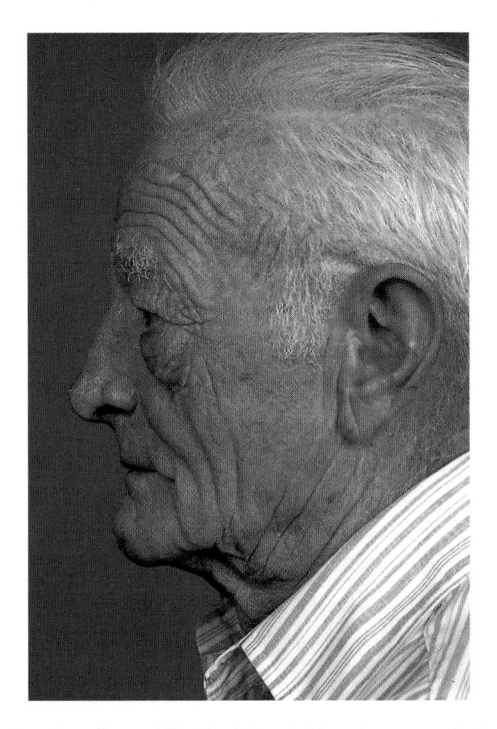

FIG. 4.3. Relaxed skin tension lines. The facial wrinkles, in general, demonstrate the lines of skin tension in the relaxed skin. Incisions made parallel to the relaxed skin tension lines (RSTL) generally heal favorably. The lines of maximal extensibility (LME) run perpendicular to the RSTL and represent the directions in which the skin can be advanced most easily. Special attention should be paid to areas such as the inferior border of the mandible where the RSTL change direction. In other areas, such as the temple and the glabella, local muscle action may be against the RSTL. For example, in the temple, the RSTL run approximately vertically rather than horizontally with the "crow's feet."

CHAPTER 5

Superficial Musculoaponeurotic System

The superficial musculoaponeurotic system (SMAS) is a continuous fibromuscular layer investing and interlinking the muscles of facial expression. The SMAS provides a conceptual framework to understand the fascial layers of the face. This relatively new concept was established by Tessier and defined by Mitz and Peyronie in 1976. The term "musculoaponeurotic" was used because of the occasional muscular fibers seen in the fascia over the parotid. The SMAS divides the subcutaneous fat into two layers. It contains fibrous septae that extend through the fat and attach to the overlying dermis (Fig. 5.1). Thus, the SMAS acts as a network to distribute facial muscle contractions to the skin.

The major vessels and nerves have consistent relationships to the SMAS within each region of the face. An understanding of these relationships can help the surgeon to protect key structures and to delineate correct dissection planes for various flaps. The incorporation of the SMAS in modern techniques for facial rejuvenation has led to longer lasting, more anatomic solutions to the problems of the aging face.

DEFINITION OF THE SMAS

The characteristics used by Tessier and his students to describe the SMAS are as follows: (1) it divides the subcutaneous fat into two layers; (2) fibrous septae extend from the dermis to the SMAS; (3) fat without septae lies between the deep facial muscles and the SMAS; (4) major vessels and nerves are initially deep to the SMAS and smaller branches perforate it, whereas the subdermal plexus is superficial to it; (5) the network of the SMAS acts as a distributor of force for the various facial muscles. This concept has held up well with the only major area of disagreement centering on the definition of the SMAS in the cheek area. In the original definition, the SMAS below the zygoma corresponds to the superficial fascia of classical anatomy and thus is superficial to the parotid fascia. Jost and Levet (1984) described cadaver and clinical studies to support their view that the true SMAS layer in the cheek includes the parotid fascia. These authors think that the parotid fascia as a remnant of the primitive platysma represents the structurally important layer of the SMAS in this area. They describe it as continuous with the platysma below and extending to the zygoma above.

Part of the complexity of these fascial layers is due to the difference in embryological development between the muscles of the lower face and those of the mid and upper face (Fig. 5.2). The muscles of the lower face and the neck derive from the embryonic primitive platysma; these include the true platysma, the risorius, the depressor anguli oris and the posterior auricular muscles, and the parotid fascia. The muscles of the mid and upper face derive from the embryonic sphincter colli profundus muscle; these include the helmet-like muscles of the skull and the mask-like muscles grouped around the orbit and the upper mouth. Of interest, the muscles derived from the primitive platysma frequently lack bony insertions, whereas those derived from the sphincter colli profundus have direct bony insertions.

The SMAS above the zygoma is a more substantial layer than the fragile fibrous network seen in the midface. There is a discontinuity in the SMAS at the level of the zygoma due to attachments of the various fascial layers at the arch. In the scalp, the SMAS is represented by the fibrous galea aponeurotica, which then splits to ensheathe the frontalis, the occipitalis, the procerus, and some of the periauricular muscles (Fig. 5.3). In the temporal region, the SMAS, the superficial temporalis fascia, and the temporoparietal fascia are synonymous.

RELATIONSHIP OF THE SMAS TO KEY NEUROVASCULAR STRUCTURES

There are important regional variations in the relationship of the SMAS to key neurovascular structures (Fig. 5.4). In the lower face, the facial nerve branches are always deep to the SMAS and innervate the facial muscles on their undersurface. The vessels and the sensory nerves in the lower face similarly arise deep to the SMAS and remain at that level except for their terminal branches. These structures are thus protected if dissection is superficial to the SMAS.

In the temporal area, the temporal branch of the facial nerve crosses the superficial aspect of the zygomatic arch and then courses within the SMAS (temporoparietal fascia) to its entrance to the frontalis muscle. In the upper face, the vessels and the sensory nerves arise from their origins (bony foramina) and penetrate the SMAS. Here they run within its superficial aspects or on its surface. Examples are the supraorbital and the supratrochlear neurovascular bundles that arise from their respective foramina, penetrate the SMAS, and then course along the superficial aspect of the frontalis muscle. Similarly, the superficial temporal vessels course with the auriculotemporal nerve along the SMAS (superficial temporalis fascia) for a few centimeters before entering the subcutaneous fat.

SURGICAL SIGNIFICANCE OF THE SMAS

The relationship of the SMAS (superficial fascia) to the deep muscular fascia can be a helpful guide to the level of important neurovascular structures. In the upper face, an understanding of the SMAS relationships is of primary importance. For example, when elevating a forehead flap, a safe and avascular plane can be found by undermining beneath the SMAS and above the periosteum. In this procedure, injury to the temporal branch of the facial nerve can be avoided by remaining beneath the SMAS and directly on the deep temporalis fascia.

As noted above, the temporal branch of the facial nerve runs within the SMAS, superficial to the deep temporalis fascia. The deep temporalis fascia consists of two layers: superficial and deep; the temporal fat pad above the zygoma is contained between the two layers. Thus, a safe plane of dissection exists through the temporal fat pad when approaching the zygoma from above (Fig. 5.3).

In the lower face the SMAS (essentially the superficial fascia) acts as a network to interconnect the facial muscles and the dermis. Its major current surgical importance for this region is as an adjunct in aging face surgery and surgery for facial paralysis, although it can be helpful in determining the level of dissection and protecting neurovascular structures. A SMAS flap partially elevated for a face-lift procedure is seen in Figure 5.5.

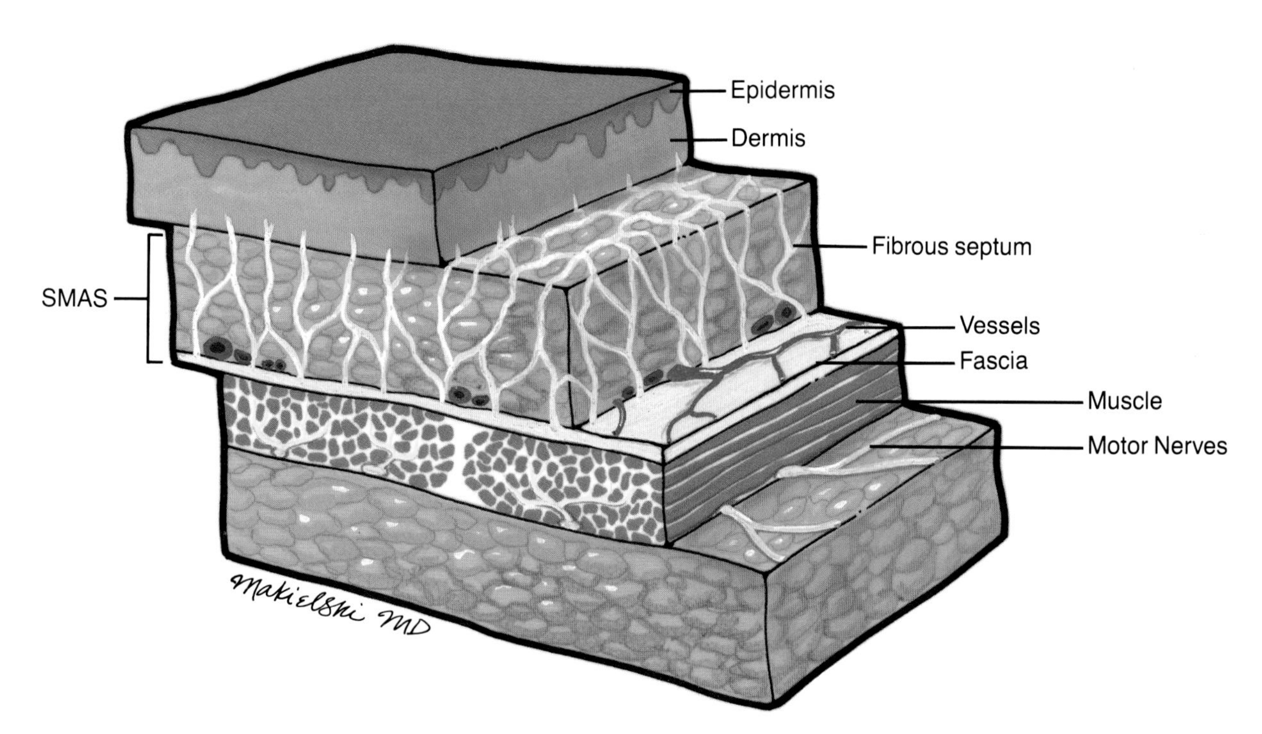

FIG. 5.1. Cross-section of the SMAS in the lower face. The SMAS envelops the facial musculature in a fibrous sheath of varying degrees of thickness and thus connects the muscles for coordination of facial expression. Fibrous septae to the dermis create a meshwork connecting the skin to the underlying muscles of expression. Motor nerves enter the deep surface of the muscles. Sensory nerves and vessels originate deep to the fascia, but their terminal branches run within or superficial to it.

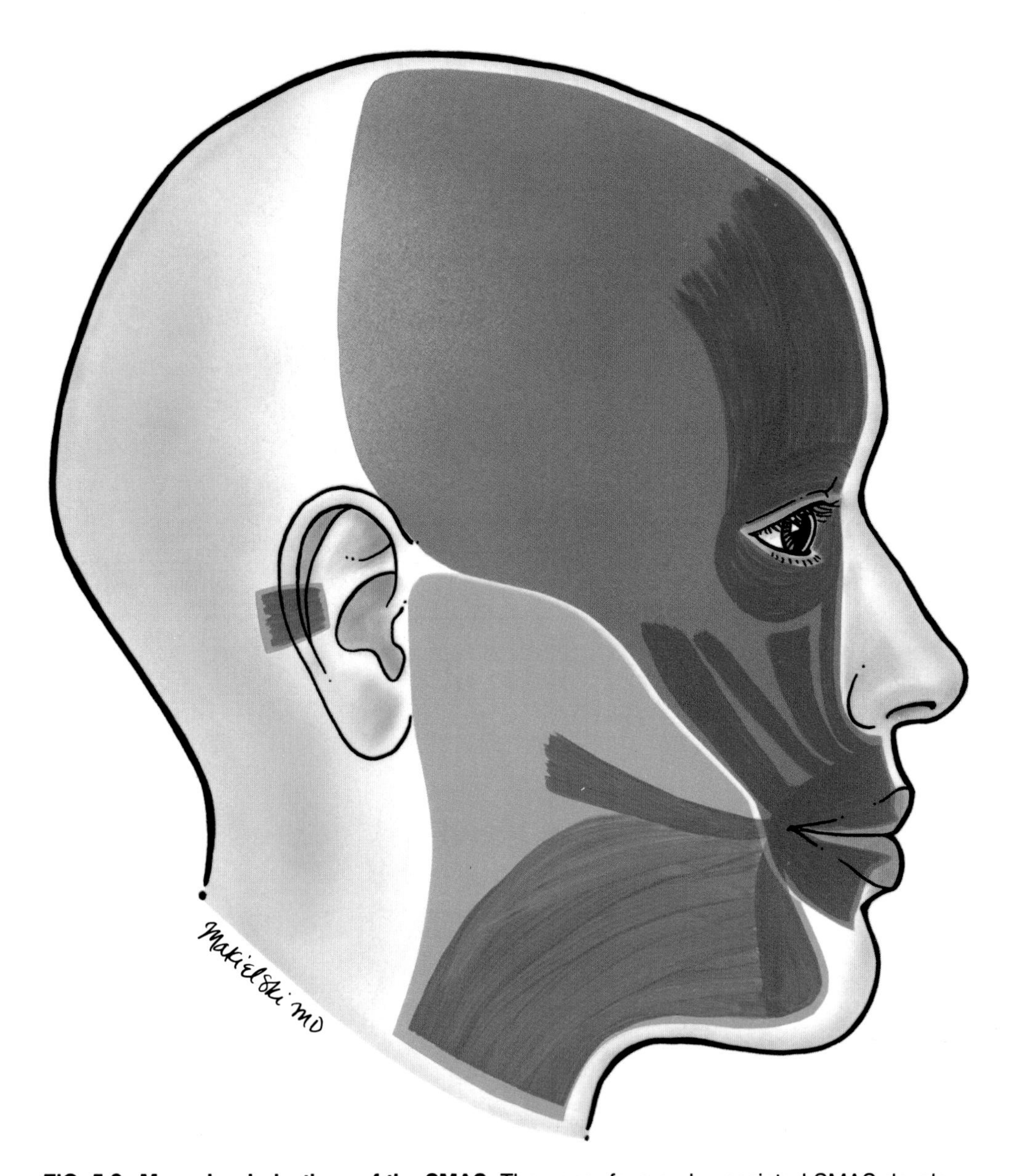

FIG. 5.2. Muscular derivations of the SMAS. The upper face and associated SMAS develop from the sphincter colli profundus. The SMAS in the lower face is derived from the primitive platysma. The frontalis m., the superficial temporal fascia, the orbicularis oculi m., the elevators of the lip, and the orbicularis oris m. constitute the upper division. The lower division includes the true platysma and its fascia, the risorius m., the depressor anguli oris m., and the posterior auricular m. The muscles derived from primitive platysma have limited bony insertions in contradistinction to those derived from the sphincter colli profundus.

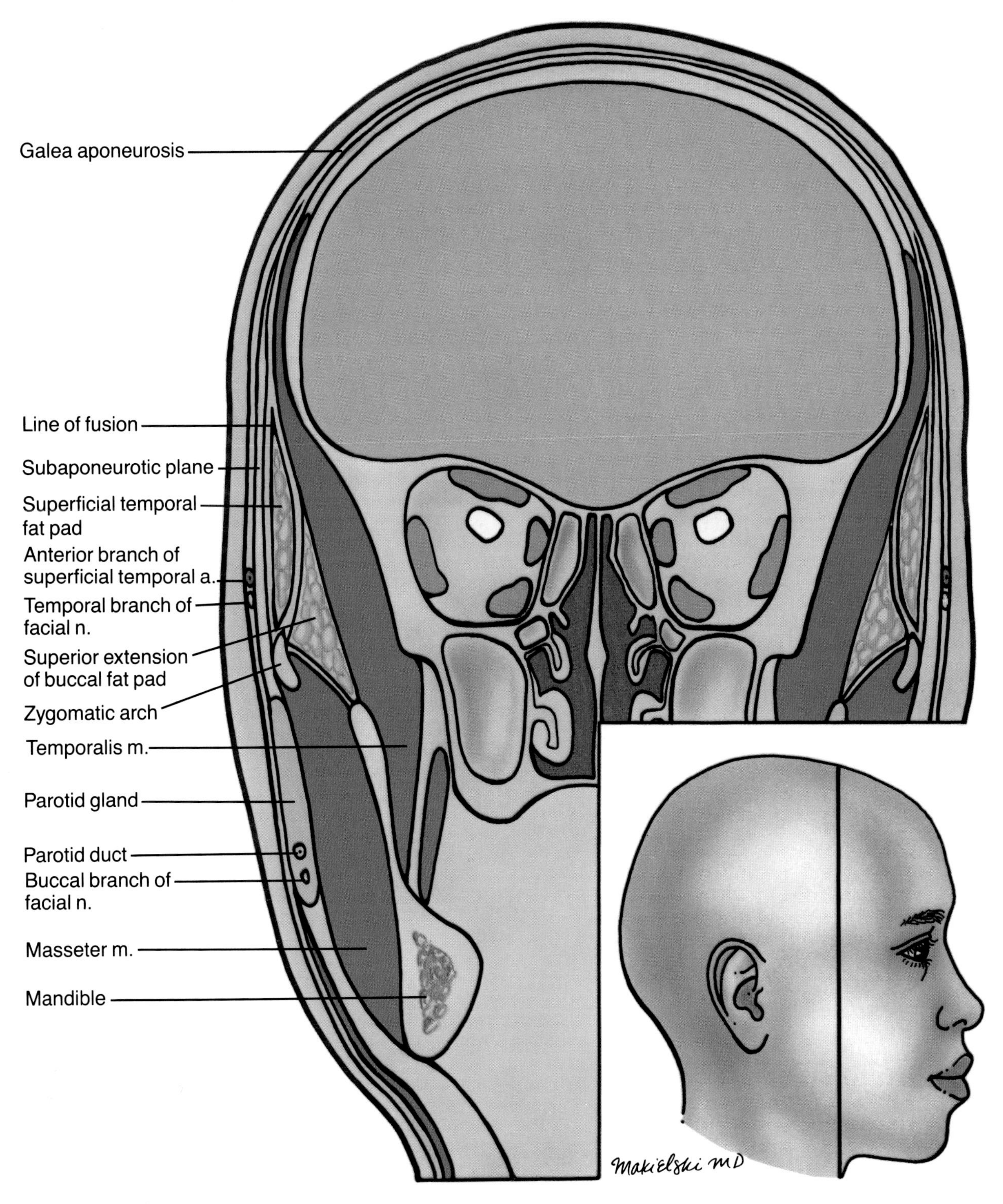

FIG. 5.3. Coronal section demonstrating the SMAS. The fibrous galea aponeurosis divides anteriorly and posteriorly to ensheath the frontalis and the occipitalis muscles. In the coronal plane, the galea aponeurosis extends to the temporalis muscle where it thins and covers the temporalis fascia as a layer of the SMAS (alternatively called the superficial temporal fascia or the temporoparietal fascia). In the temple above the zygoma, the temporal branch of the facial nerve and the anterior branch of the superficial temporal artery run within this layer of the SMAS. The deep temporalis fascia splits at its line of fusion into deep and superficial layers. Both attach to the zygoma and between them lies the superficial temporal fat pad. The superior extension of the buccal fat pad lies deep to the deep layer of the temporalis fascia. Inferior to the zygoma, the SMAS covers the parotid gland and becomes contiguous with the platysma. In this plane, the parotid duct and the buccal branch of the facial nerve lie within the parotid. The muscles of mastication surround and insert into the mandible.

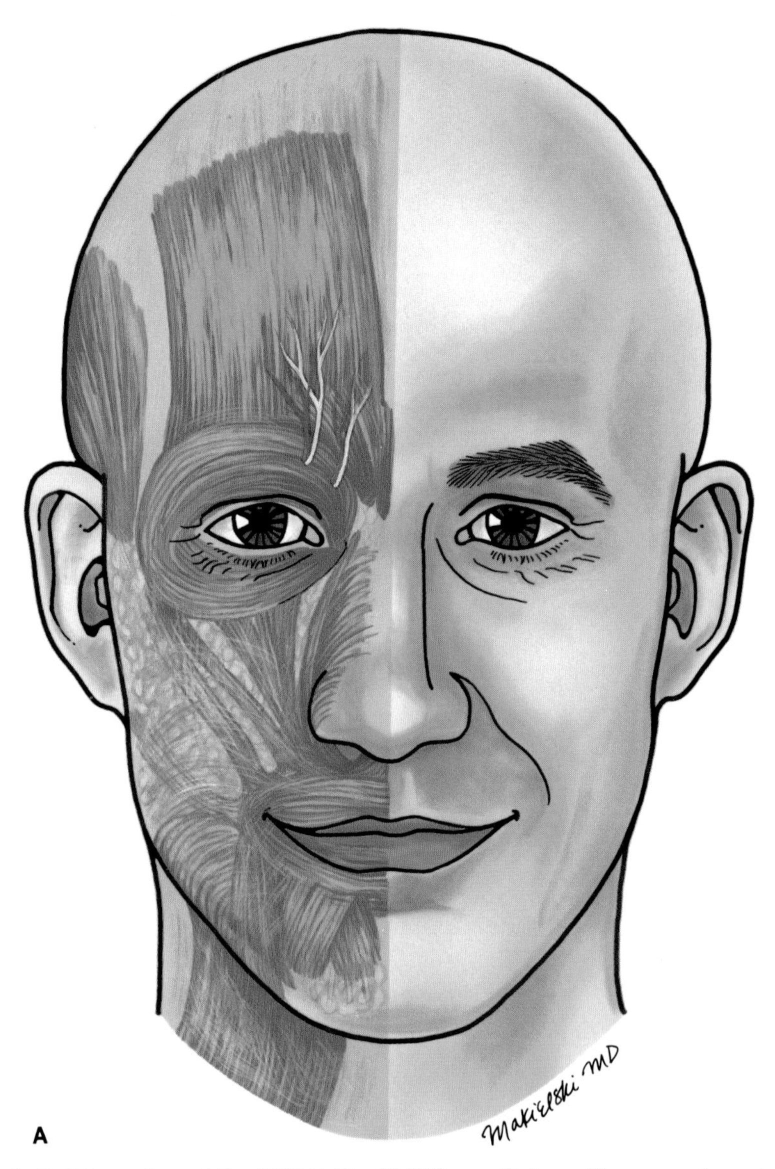

FIG. 5.4. A: Full-face view of the SMAS. The SMAS constitutes a fibromuscular network that interconnects the muscles of facial expression. In addition to its usefulness as a deep layer to tighten in aging face surgery, it serves as a guide to the depth of key neurovascular structures. In the lower face, the facial nerve branches enter the muscles of facial expression on their deep side and are always deep to the SMAS. Similarly, in the lower face, vessels and sensory nerves originate deep to the SMAS and their terminal branches run with it superficially. In the upper face, the important neurovascular structures arise from their deeper origins and course within the SMAS. The infratrochlear and supratrochlear neurovascular pedicles lie within the SMAS over the frontalis muscle.

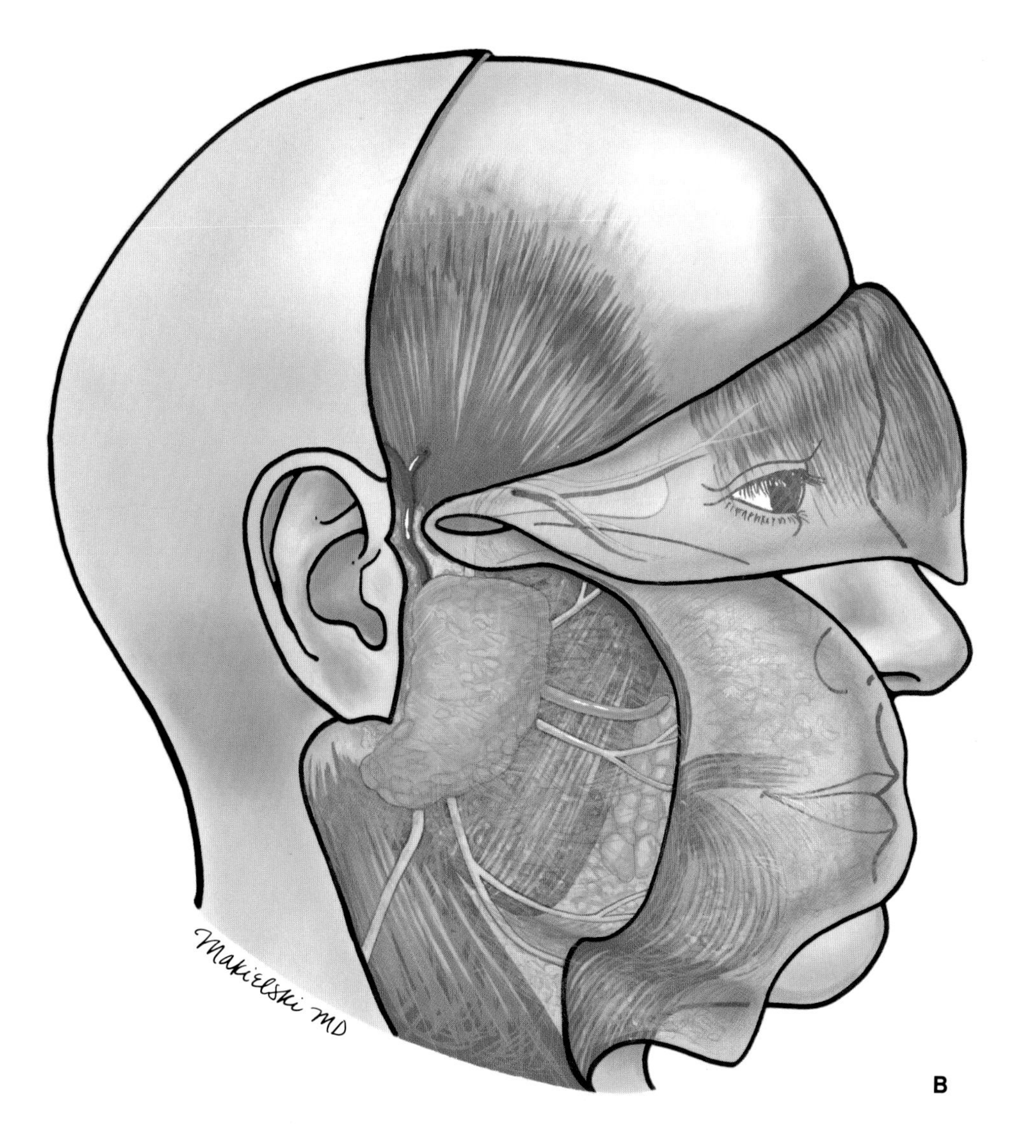

FIG. 5.4. *Continued.* **B: The SMAS reflected.** The SMAS is somewhat discontinuous at the zygoma. Below the zygoma, the SMAS is substantial over the parotid and the neck but becomes a more tenuous "meshwork" over the anterior cheeks, the lips, and the nose. The facial nerve branches and the parotid duct are seen deep to the SMAS and superficial to the masseter and the buccal fat pad. Above the zygoma, the SMAS is a more consistent layer that envelops the frontalis muscle and then forms the galea aponeurotica. The temporal branch of the facial nerve runs with the SMAS in the temple and is particularly vulnerable over the zygoma where it lies quite superficially just beneath the subcutaneous fat.

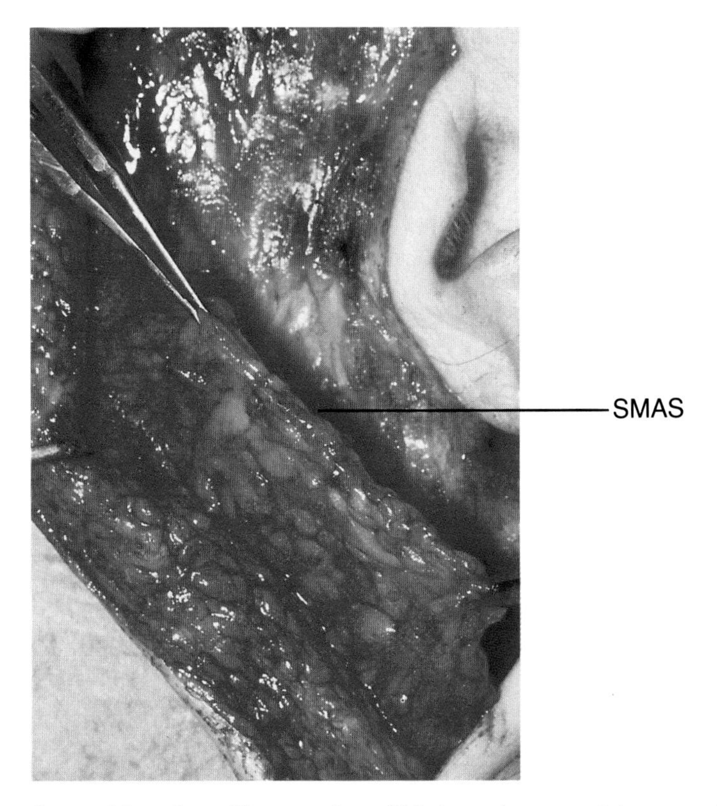

FIG. 5.5. SMAS flap elevated in a face-lift procedure. This layer forms an interconnecting meshwork to support the facial muscles and associated soft tissues. The SMAS over the parotid is lifted in continuity with the platysma to achieve a deep layer closure in the face-lift procedure. The facial nerve branches are deep until the anterior border of the parotid is reached.

CHAPTER 6

Facial Musculature

The facial musculature of interest to the facial surgeon can be divided into two groups: the muscles of facial expression and the muscles of mastication. Figure 6.1 shows the major muscle attachments on the skull. The major muscle groups are shown in Figure 6.2.

MUSCLES OF FACIAL EXPRESSION

The muscles of facial expression, or mimetic muscles, move the skin of the face and the scalp and act as sphincters for the eyes, the nose, and the mouth. These muscles develop from the mesoderm of the second branchial arch, the hyoid arch. In addition to the scalp, the facial, and the platysma muscles, three deeper muscles also develop from the hyoid arch: the stapedius, the stylohyoid, and the posterior belly of the digastric muscle. The muscles of facial expression are thin, flat muscles innervated by the facial nerve, which is the motor nerve of the second branchial arch. There is considerable individual variation and frequently these muscles blend into one another.

Forehead

The frontalis and the corrugator muscles are the primary muscles of the forehead. The frontalis muscle courses vertically and inserts into the galea aponeurotica (Fig. 6.3). It raises the eyebrows and produces transverse forehead wrinkles. A paralysis of the temporal branch of the facial nerve, which innervates the frontalis muscle, causes a unilateral brow ptosis and loss of forehead motion. The corrugators (Fig. 6.4) arise from the medial orbital rim just above the nose and insert into the frontalis muscle and the skin of the eyebrows. These paired muscles act to pull the eyebrows together and to produce glabellar frown lines. The corrugator, the procerus, and the obicularis oculi muscles act in concert to close the eyes and create transverse and oblique glabellar lines.

The frontalis muscle forms the frontal belly of the occipitofrontal muscle. The fibers of the frontalis course vertically and insert into the galea aponeurotica on the top of the head (Fig. 11.3). The frontal belly pulls against the occipital belly through the galea when raising the eyebrows.

Eye

The orbicularis oculi surrounds the orbit and extends into both eyelids. It arises from the medial palpebral ligament and is responsible for eye closure and blinking. The orbicularis oculi muscle has two parts: the palpebral, and the orbital (Fig. 6.5). These relationships are described in Chapter 13.

Ear

The auricle has three extrinsic muscles: the anterior, the superior, and the posterior auricular muscles (Fig. 6.6). In general, they are poorly developed and not clinically significant.

Nose

The nasal muscles are the procerus, the nasalis, and the depressor septi (Fig. 6.7). The procerus is the most superior of this group, arising from the nasal bone in the glabellar region and inserting into the forehead skin. The procerus pulls forehead skin down and may cause transverse creases between the brows.

There are two parts to the nasalis muscle. The upper, transverse part attaches to its counterpart across the dorsum of the nose and acts to compress the naris, hence its other name, compressor naris. The second part is the alar portion, or dilator naris, so named because it elevates the lower lateral cartilage to flare the nostril.

Cheek

The buccinator muscle (bugler's muscle) arises posteriorly from the pterygomandibular raphe, deep to the buccal fat pad, and extends anteriorly to attach to the mouth at the orbicularis oris muscle (Fig. 6.2). The buccinator functions to keep food between the teeth during chewing and maintains air pressure during blowing.

Mouth and Lips

The mouth is encircled by fibers of the orbicularis oris, which functions as a sphincter (Fig. 6.8). Various muscles insert into the orbicularis and elevate or depress the lips and open the mouth (Figs. 6.9 and 6.10).

The lip elevators include the major and the minor zygomatic, the levator labii, and the levator anguli oris. The corner of the mouth is moved by the major zygomatic, the risorius, and the levator anguli oris.

The lip depressors arise from the lower border of the mandible and include the depressor anguli oris and depressor labii (Fig. 6.11). The mentalis muscle protrudes the lower lip.

MUSCLES OF MASTICATION

Among the muscles of mastication, the masseter and the temporalis muscles have the greatest surgical relevance; both are useful for reconstructive procedures. The pterygoid muscles are of less significance in facial surgery.

The Masseter

The masseter muscle arises in two portions from the zygomatic arch. Its superficial part arises from the lower border of the anterior two-thirds of the arch, whereas its deep part arises from the inner surface of the arch and its posterior third (Fig. 6.12). The fibers of the superficial part run inferiorly and posteriorly, whereas the deep fibers run almost directly inferiorly. The muscle as a whole inserts into the entire lateral ramus of the mandible. The nerve supply is the masseteric from the mandibular division of V, and the arterial supply is the masseteric from the maxillary artery. The neurovascular bundle runs through the coronoid notch to enter the deep surface of the muscle, where it arborizes in an oblique anteroinferior direction (Correia and Zani, 1973).

The parotid gland overlies the posterior portion of the masseter, and the buccal branches of the facial nerve are closely applied to the anterior muscle body by the parotid-masseteric fascia (Fig. 6.13). The buccal fat pad lies just deep to its anterior edge.

Rotation of the masseter muscle is a proven method of facial reanimation in cases of longstanding facial paralysis (Conley, 1975). Through a parotidectomy-type incision, the inferior border is incised and the muscle is freed from its attachments to the mandibular ramus. It is then split into two or more strips that are sewn to the superior and the inferior orbicularis oris through subcutaneous tunnels (Fig. 6.14). When the patient smiles, clenching the teeth will direct the pull of the masseter to the corner of the mouth, producing a symmetric voluntary smile. Injury to the masseteric nerve will obviate this result. The nerve is less likely to be injured if the incision dividing the muscle is made as posteriorly as possible and if the division of the transposed muscle into strips is made in the axis of the nerve, i.e., in an oblique anteroinferior direction. The temporalis muscle is frequently transposed simultaneously to give some superior direction to the muscular forces.

The Temporalis

The temporalis is a broad, fan-shaped muscle originating from the temporal fossa on the side of the skull (Fig. 6.15). Its fibers converge into a tendon that inserts into both the medial side of the ramus of the mandible and the entire coronoid process. The temporal extension of the buccal fat pad lies on the temporalis tendon, separating it from the zygomatic arch, and then continues around the anterior edge of the muscle along the lateral orbital wall. The innervation of the temporalis is from the mandibular branch of the trigeminal nerve (V3) via two (usually) deep temporal nerves, the anterior and the posterior. The blood supply is from the temporal branches of the maxillary artery. The blood vessels and the nerves travel together from the infratemporal fossa to enter the deep surface of the temporalis.

The temporalis is useful in facial reanimation procedures. When freed from its superior cranial periosteal attachments, it can be turned over the zygomatic arch to tighten the corner of the eye, or, with strips of deep temporalis fascia to lengthen it, it can reach the corner of the mouth. Tunneled under the zygomatic arch, it can be turned as a muscle flap into the oral cavity to reconstruct the palate and the tonsillar areas after radical resections.

Direction of Muscle Pull on the Mandible and the Zygoma

Fractures of the zygoma or the mandible result in displacement of the free segment by the attached muscles, namely, the masseter, the temporalis, the medial, and the lateral pterygoid, and the muscles of the floor of the mouth. The direction of the displacement is predicted by the direction of pull of these muscles (Figs. 6.12, 6.16, and 6.17). Fractures in which the muscle tension tends to stabilize the fracture are termed "favorable"; those in which the muscle tension tends to displace the free segment are termed "unfavorable." Knowledge of these directions, therefore, assists in determining the appropriate management.

The Temporalis Fascia

The temporalis is covered by a dense, tough fascia, the deep temporalis fascia, which is continuous with the periosteum of the skull. Some fibers of the temporalis arise from this fascia. A few centimeters above the zygomatic arch, at the temporal line of fusion, the deep temporalis fascia splits into its superficial and its deep layers, which then insert into the superficial and the deep aspects of the superior surface of the zygomatic arch. Between these two layers is found the wedge-shaped, superficial temporal fat pad, which sits on the superior surface of the zygomatic arch. The middle temporal artery, which arises from the superficial temporal artery at the level of the zygomatic arch, enters the deep temporalis fascia immediately above the zygomatic arch, supplies the fascia and the superficial temporal fat pad, and gives off branches to the temporalis itself.

Superficial to the deep temporalis fascia is another fascial layer, the superficial temporalis fascia. It lies immediately deep to the dermis and is continuous with the galea aponeurotica above and the SMAS of the face below. The superficial temporal artery and its anterior and its posterior branches lie within the fascia, and the accompanying vein lies immediately superficial to it. On its superficial surface, it is attached somewhat to the subdermal tissues, whereas on its deep aspect, there is an avascular plane separating it from the deeper structures. The temporal branch of the facial nerve lies on the deep aspect of the fascia, 1 to 2 cm posterior to the lateral orbital rim. The surgeon is more likely to recognize the superficial temporalis fascia after passing through it, thus putting the nerve at risk. A surgical approach that goes deep to the superficial layer of the deep temporalis fascia inferior to the line of fusion will avoid the nerve.

Both the deep and the superficial temporalis fascial layers have axial blood supplies, provided by the middle temporal artery and the superficial temporal artery, respectively. This axial supply allows these layers to be harvested and transferred as microvascular free flaps to other areas of the body. If the superficial temporal artery is taken proximal to the origin of the middle temporal artery, both fascial layers may be harvested as a bilobed free fascial flap.

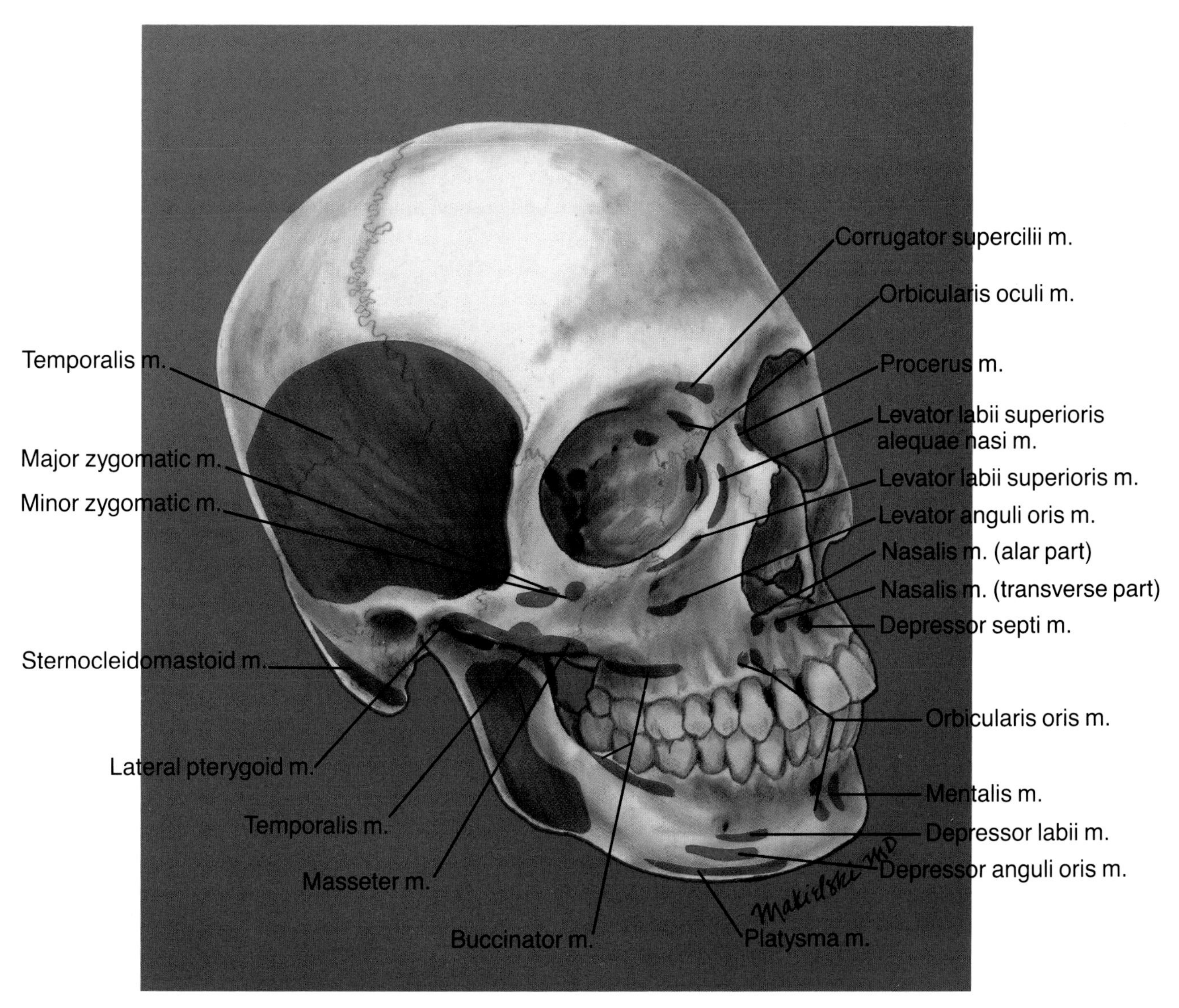

FIG. 6.1. Attachments of facial muscles to the skull. The muscles of the lower face derived from primitive platysma have primarily soft-tissue insertions, whereas the remainder insert into the craniofacial skeleton.

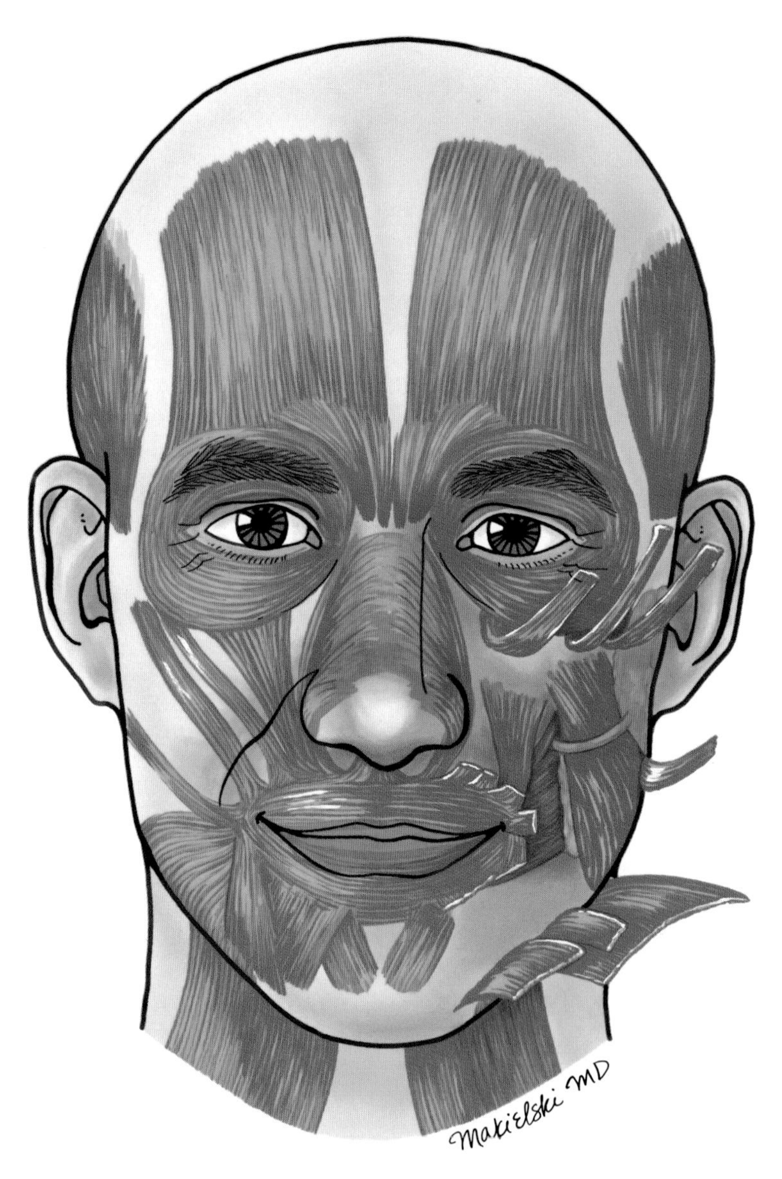

FIG. 6.2. Musculature of the face. The risorius, the major and the minor zygomatic, and the levator labii superioris muscles on the left have been reflected to show the underlying levator anguli oris and the buccinator muscles.

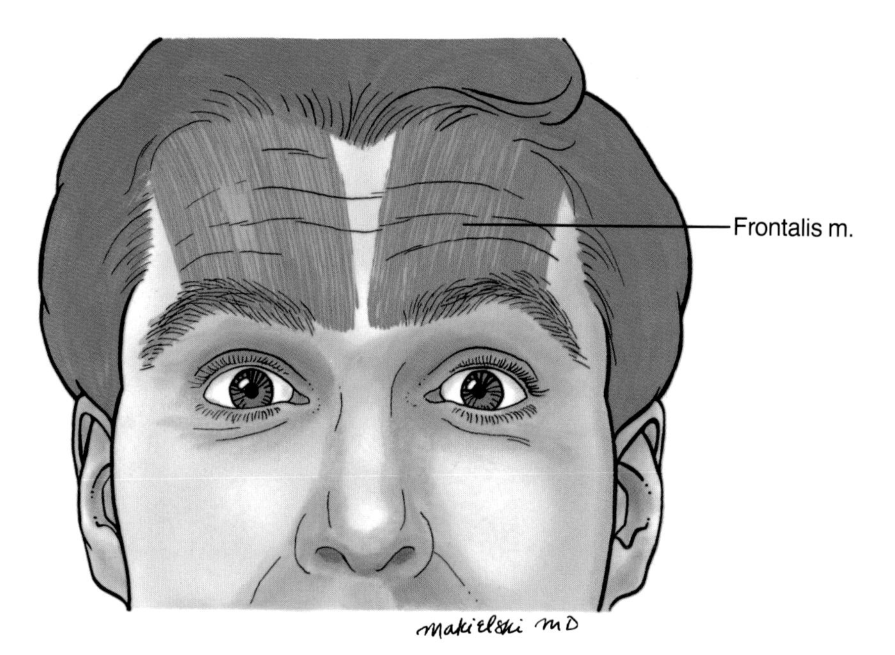

FIG. 6.3. Frontalis muscle. The frontalis muscle elevates the brows and creates the horizontal forehead creases. The lateral extent of the frontalis muscle corresponds to the extent of the forehead creases.

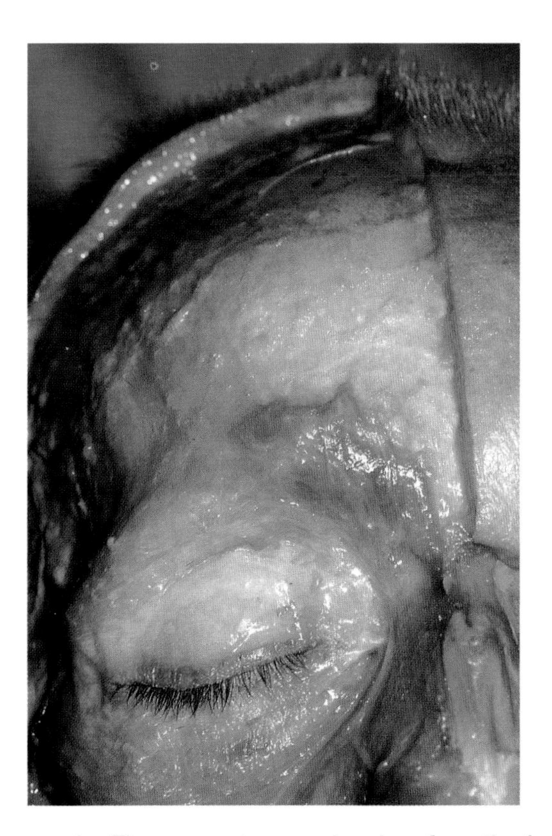

FIG. 6.4. Corrugator muscle. The corrugator muscle arises from the frontal bone near the superior orbit and inserts into the skin in and above the medial brow. Its contraction causes "glabellar frown lines."

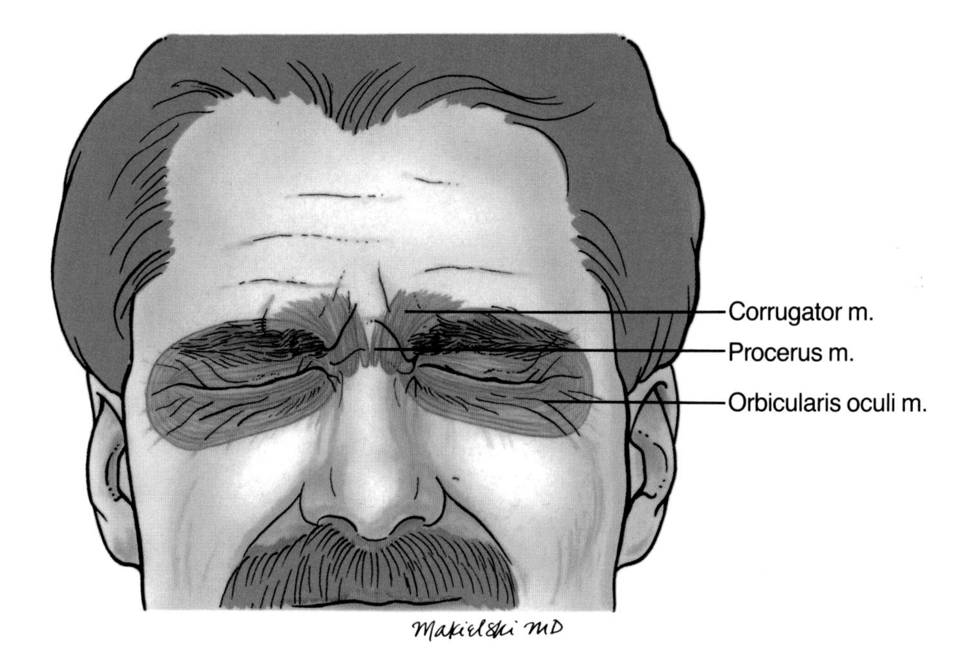

FIG. 6.5. Orbicularis oculi, procerus, and corrugator muscles. These muscles act as a unit to close the eyes and create facial expressions such as a "squint."

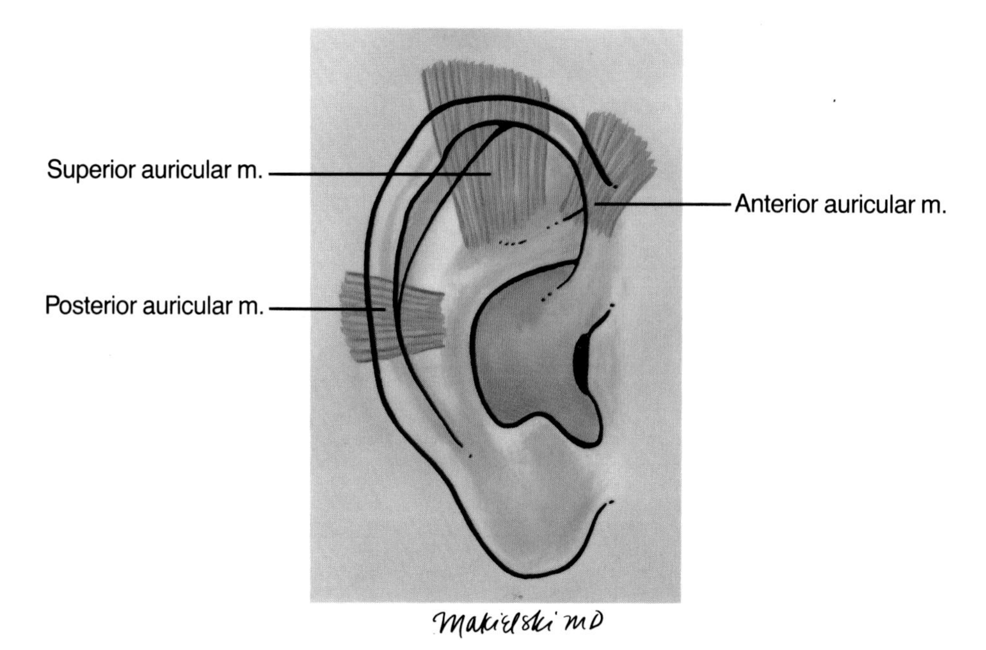

FIG. 6.6. Ear musculature. The three primary ear muscles in the human are rudimentary and provide primarily soft-tissue support.

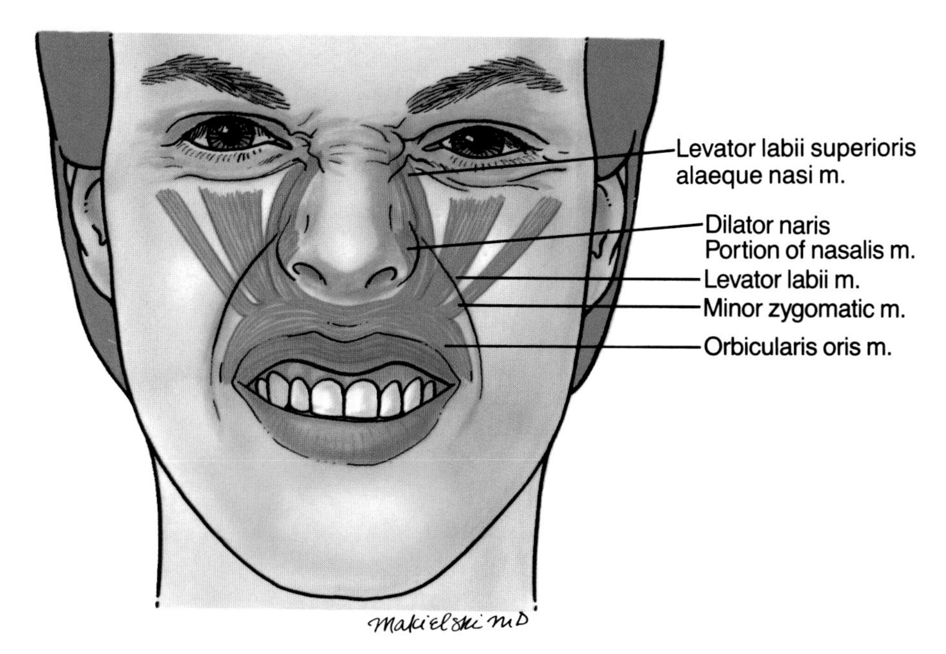

FIG. 6.7. Upper lip and nasal musculature. The orbicularis oris muscle is primarily a sphincter essential for oral competence, speech, and social expression. Its contraction is opposed superiorly by the action of the major lip elevators: the nasalis, the levator labii, and the minor zygomatic muscles.

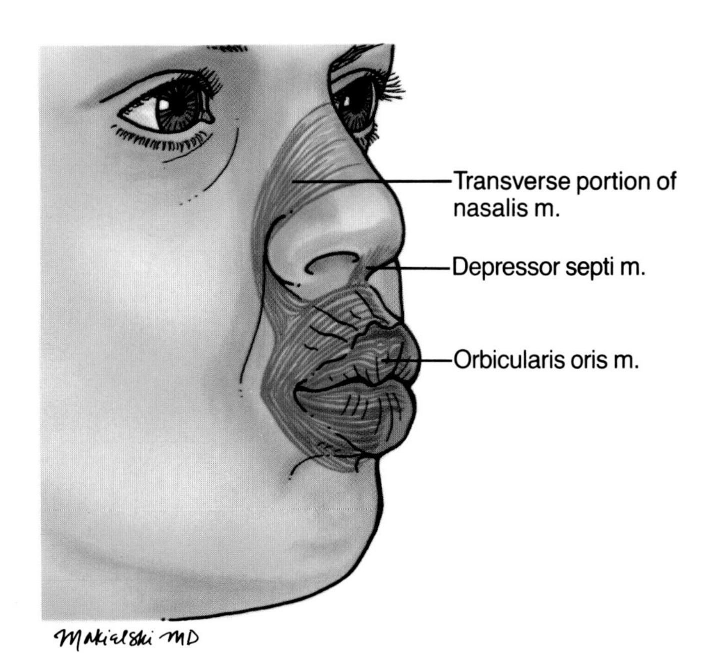

FIG. 6.8. Orbicularis oris and nasal musculature. The nasalis and the depressor septi muscles interdigitate with the orbicularis oris muscle. A strong depressor septi muscle depresses the nasal tip with smiling and can be interrupted during rhinoplasty surgery to decrease active tip depression and help to maintain tip position.

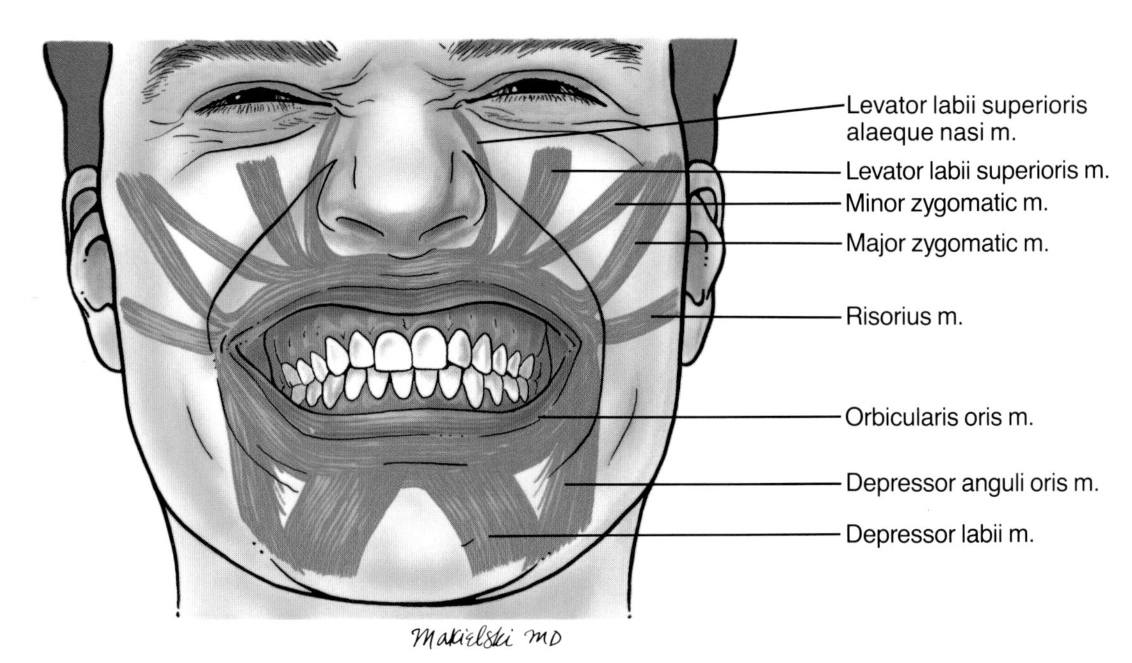

FIG. 6.9. Musculature of the mouth and the lips. The multiple functions of the lips (oral competence, speech, social expression) require a complex set of motions facilitated by the antagonism between the circular contraction of the orbicularis oris and the radial action of the lip elevators and depressors.

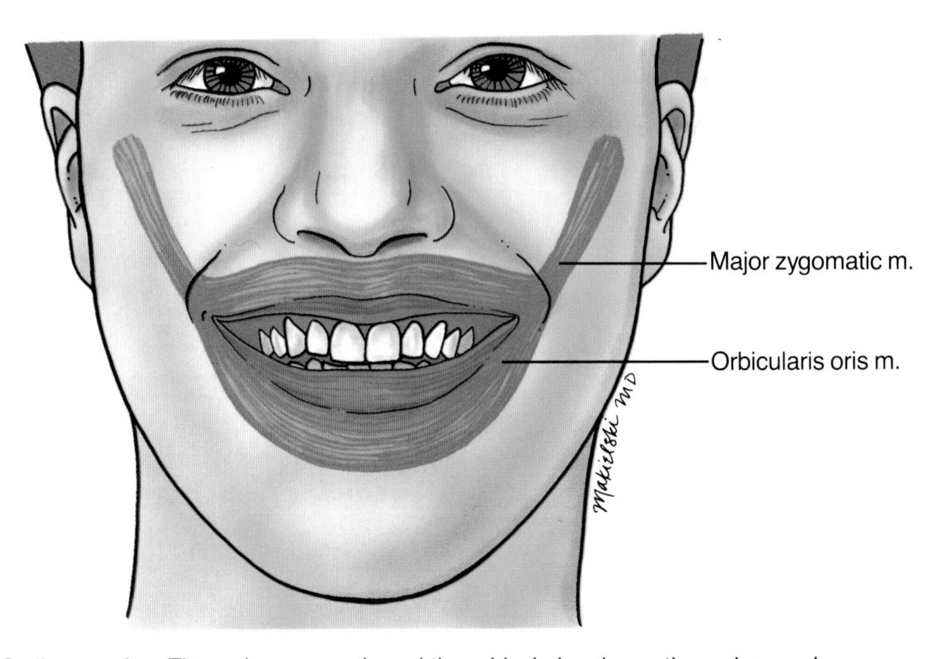

FIG. 6.10. Smile muscles. The major zygomatic and the orbicularis oris are the main muscles used to smile. Other muscles contribute to the subtle lip movements needed for the vast range of facial expression. The major zygomatic muscle contracts in approximately the same vector as that desired for the temporalis muscle when it is transferred for facial reanimation.

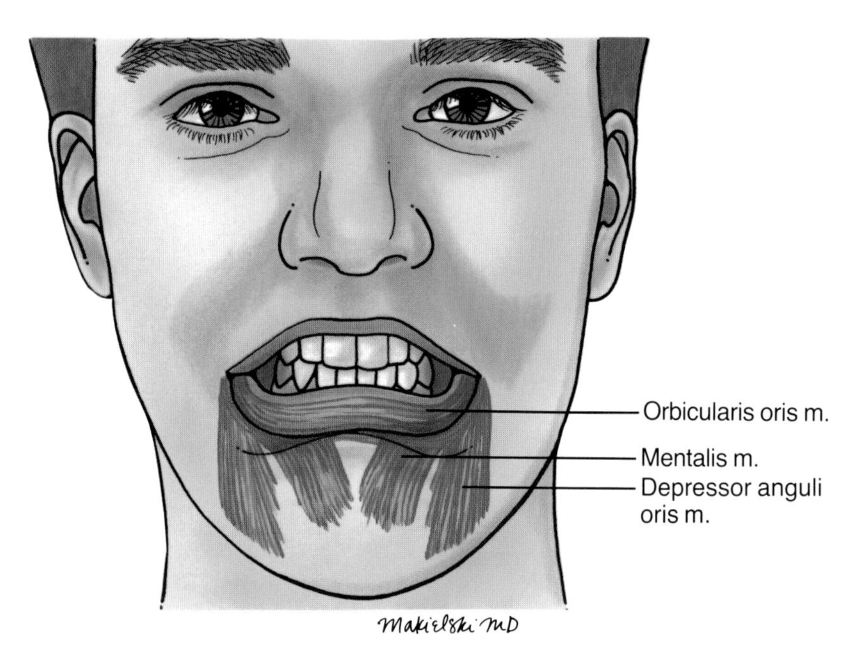

FIG. 6.11. Lip depressors. These muscles depress the lower lip. They are innervated by the marginal mandibular branch of the facial nerve. Unilateral paralysis of this branch results in the typical facial asymmetry seen with smiling, in which the counteraction of the orbicularis oris muscle creates an elevation of the lip on the paralyzed side. (Depressor labii inferioris has been removed to better show the mentalis m.)

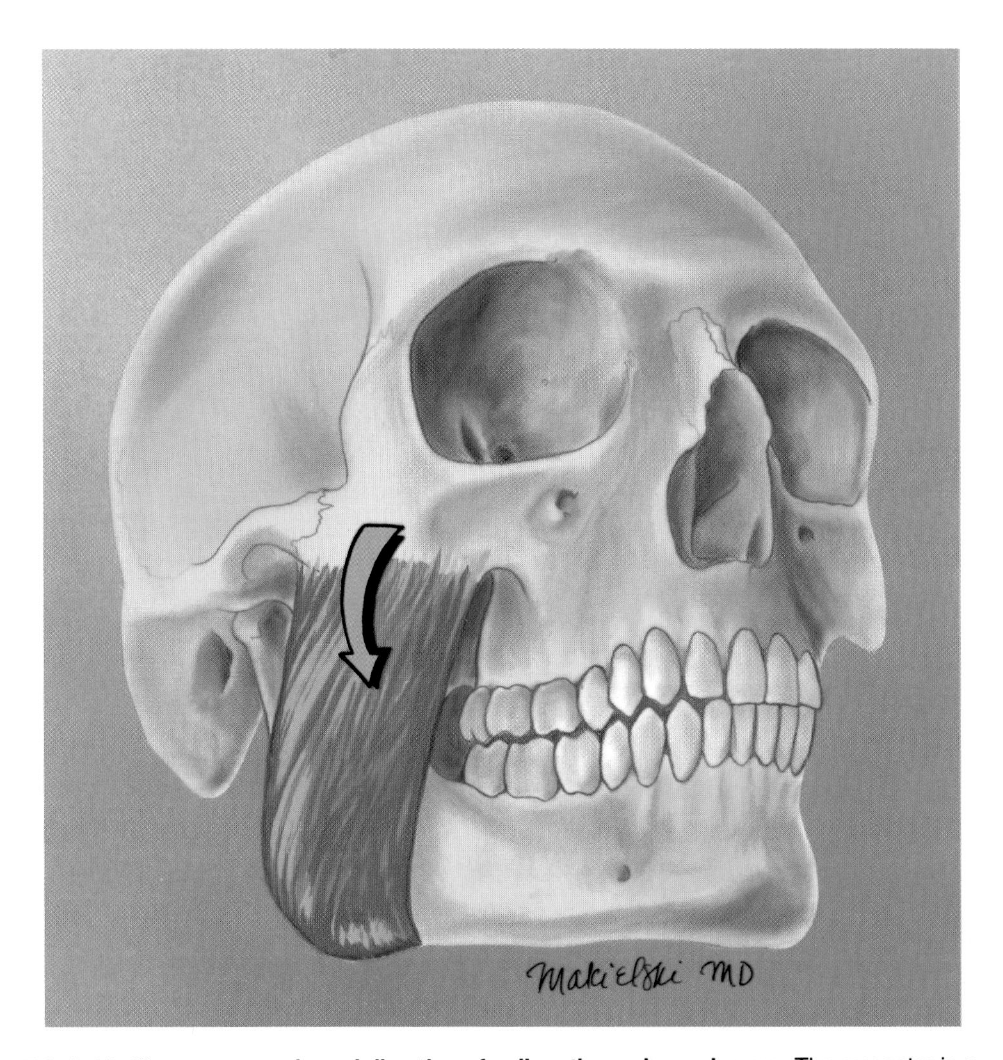

FIG. 6.12. Masseter muscle and direction of pull on the malar eminence. The masseter is a short, strong muscle that elevates the mandible. It tends to distract the malar eminence inferiorly and medially after a trimalar fracture. As one of the muscles of mastication, it is innervated by the fifth cranial nerve and thus can be used for muscle transpositions when there is paralysis of the facial nerve.

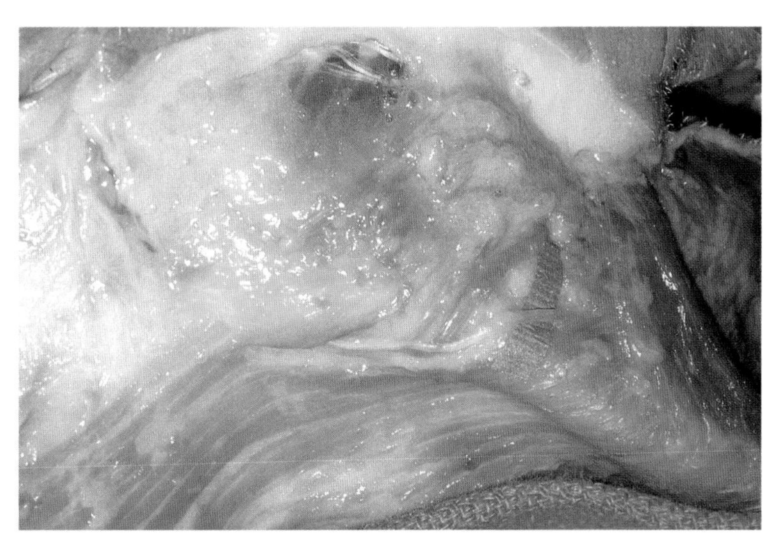

FIG. 6.13. Masseter muscle with facial nerve branches. Buccal branches of the facial nerve course over the masseter within the fascia after they exit the parotid gland.

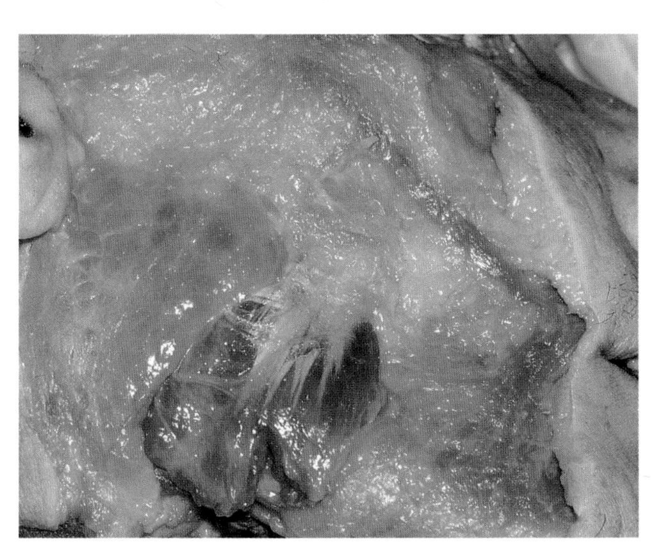

A

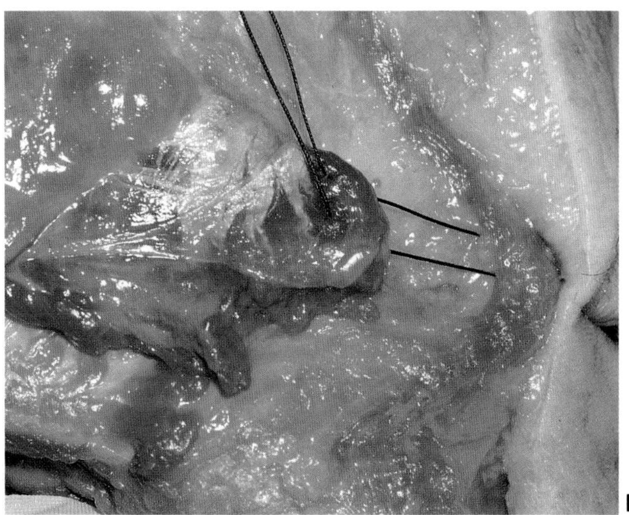

B

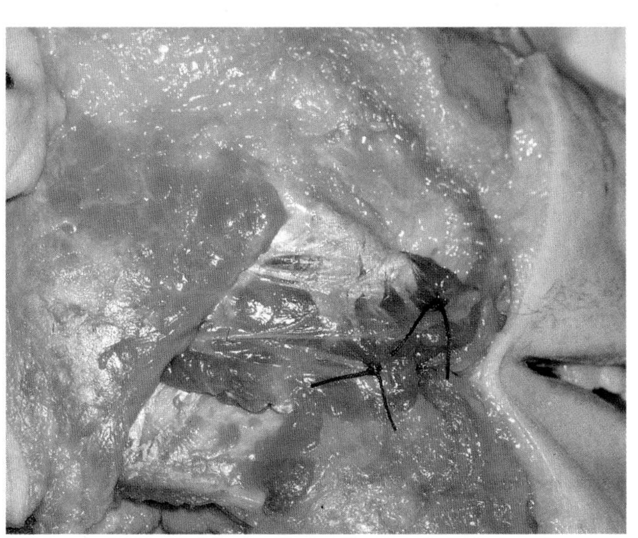

C

FIG. 6.14. Masseter muscle transfer. The masseter muscle is usually transferred as a unit to avoid injury to its motor nerve. This dissection demonstrates the principles of the transfer (Conley, 1975). The temporalis muscle can be similarly transferred. The temporalis has a more superior direction of pull and therefore gives a more natural elevation to the corner of the mouth than the masseter. **A:** The muscle belly is divided vertically to create two segments. The incision is carried no more than 1.5 cm superiorly to avoid motor nerve injury (Correia and Zani, 1973). **B:** The muscle is transposed to the upper and the lower lips. **C:** Overcorrection is performed to compensate for the short contraction of the muscle and the primarily posterior pull. A temporalis transfer is frequently combined to obtain this superior force vector.

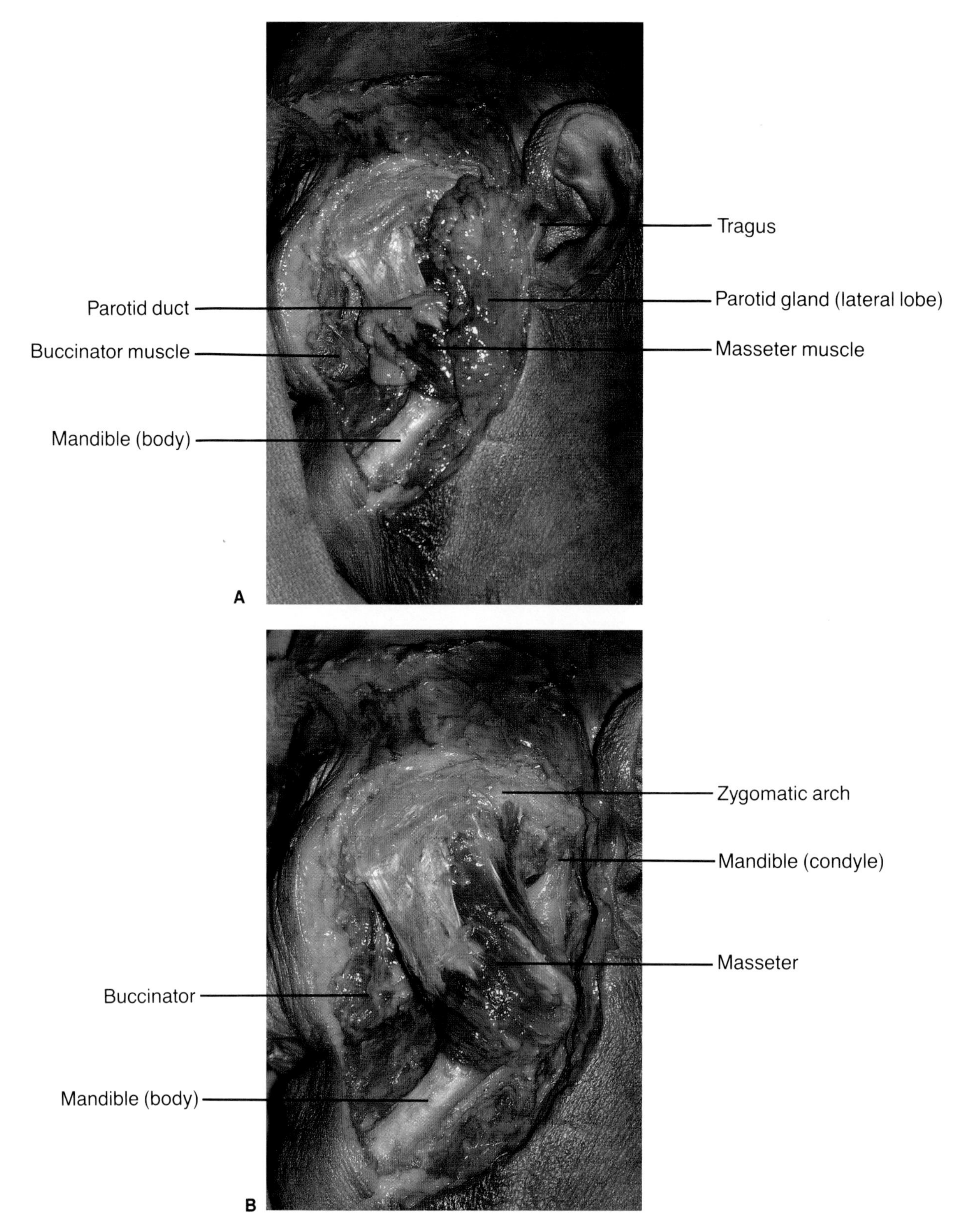

FIG. 6.15. Deep relationships of the masseter and the temporalis muscles. A: The parotid duct emerges at the anterior border of the parotid and crosses lateral to the masseter muscle. At the anterior border of the masseter, the duct turns medial and pierces the buccinator. (The facial nerve has been removed.) **B:** The lateral parotid and the facial nerves have been removed, exposing the condyle. The masseter is seen as it inserts into the ramus of the mandible.

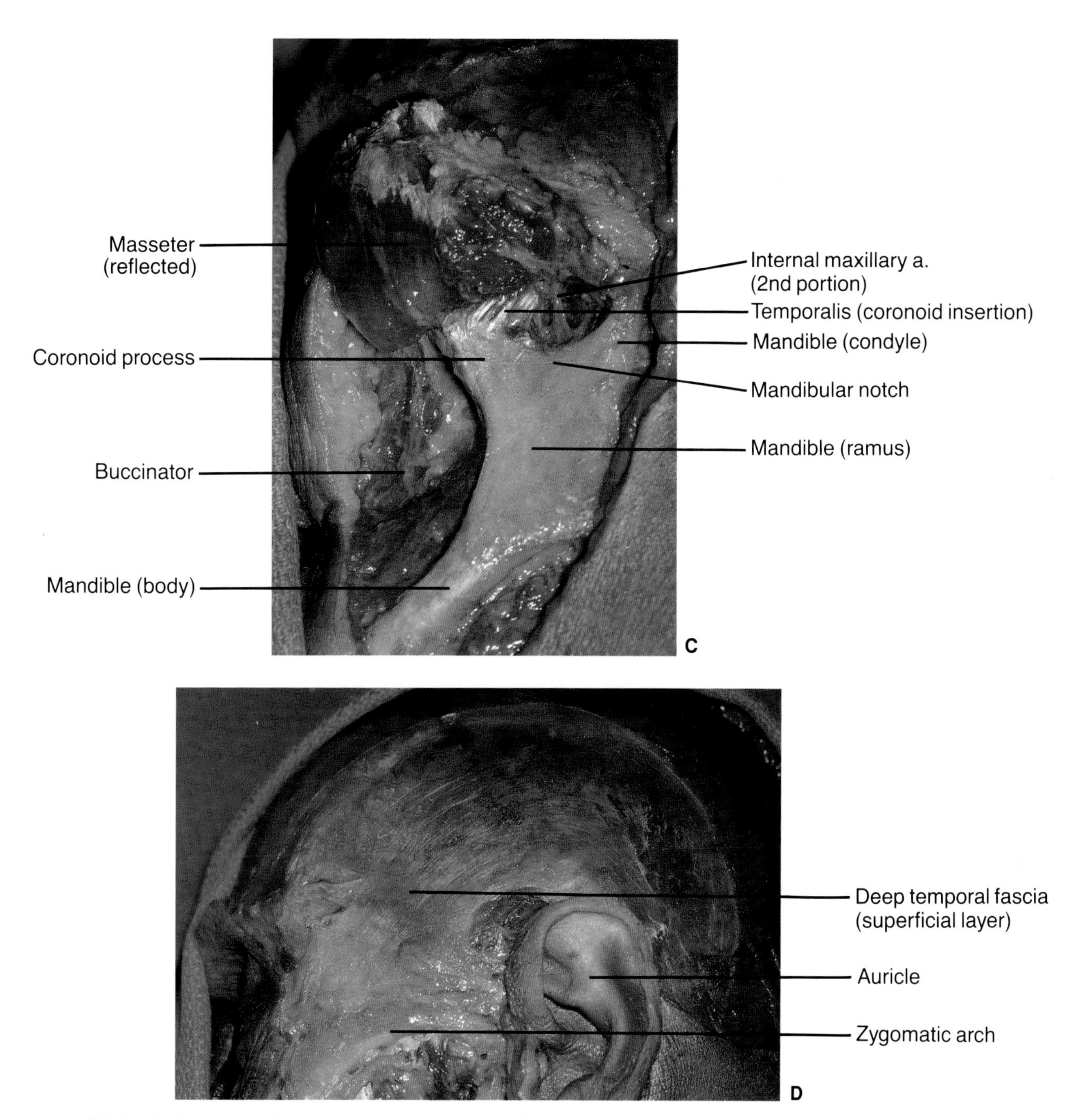

FIG. 6.15. *Continued.* **C:** The masseter has been reflected from the ramus of the mandible to show the vascular supply. The insertion of the temporalis muscle into the coronoid process is demonstrated, and the contents in the mandibular notch are clearly visible. **D:** The superficial layer of the deep temporal fascia. The superficial temporal fascia has been removed.

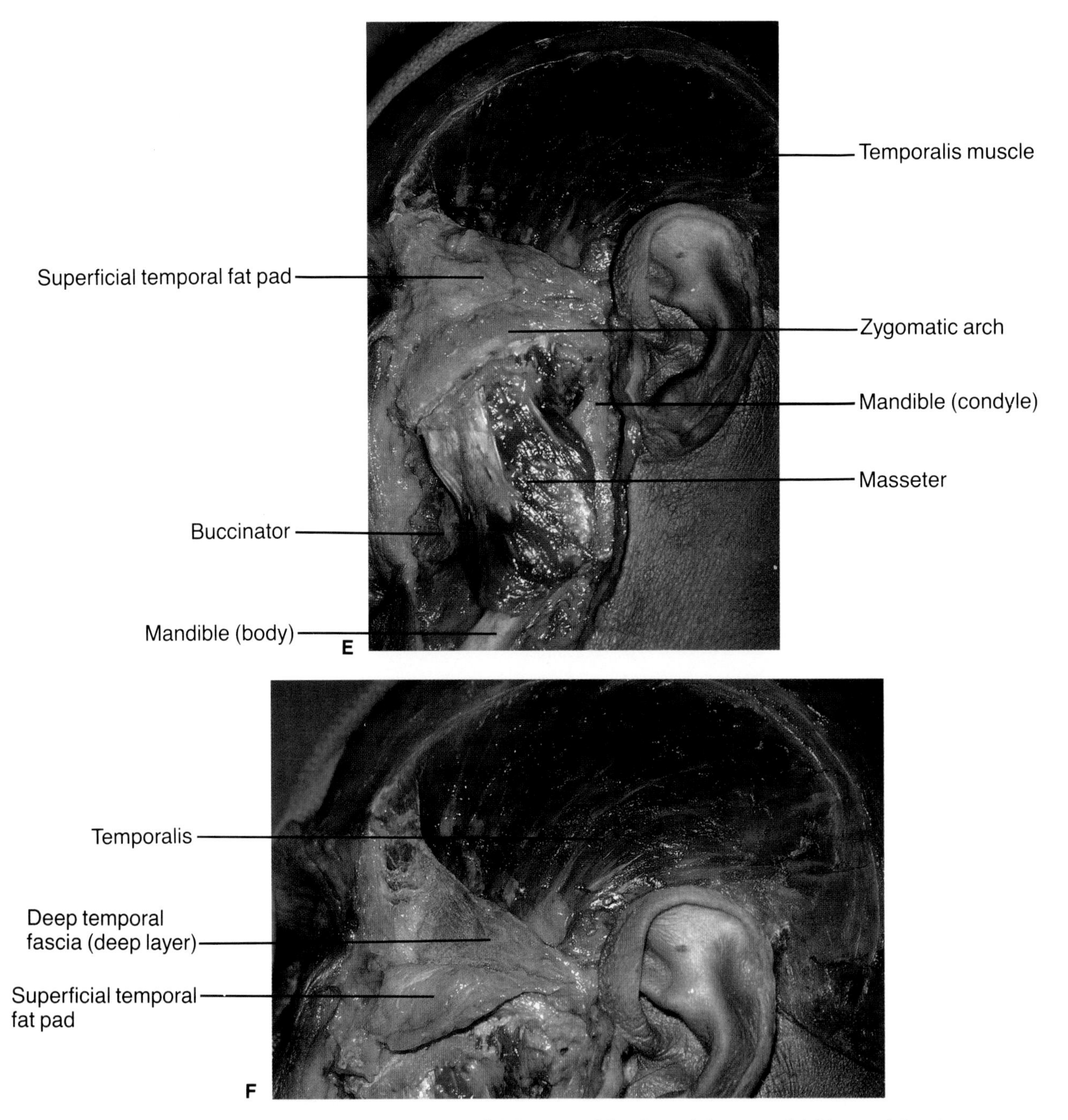

FIG. 6.15. *Continued.* **E:** The superficial temporal fascia and the superficial layer of the deep temporal fascia have been removed to the line of fusion, revealing the superficial temporal fat pad. **F:** The superficial layer of the deep temporal fascia has been removed. The deep layer of the deep temporal fascia has been preserved and is shown underlying the superficial temporal fat pad.

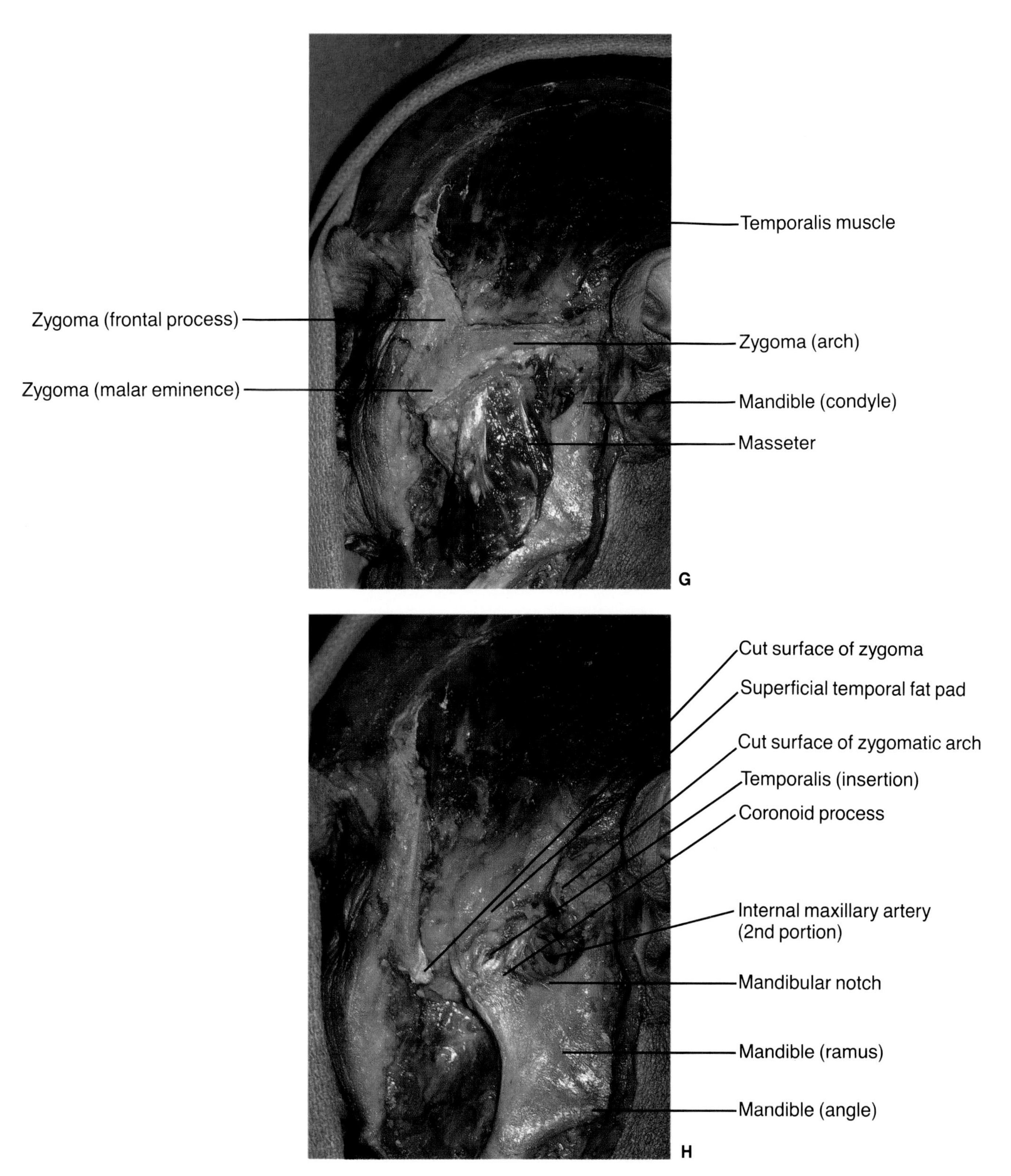

FIG. 6.15. *Continued.* **G:** The superficial temporal fat pad has been removed. The close relationship between the zygoma, the masseter, and the temporalis is clearly seen. The masseter has been partly reflected along its insertion into the ramus. **H:** The deep layer of the lateral face. The lateral parotid, the facial nerve, the zygomatic arch, and the masseter have been removed. A branch of the maxillary artery is seen deep in the mandibular notch. The superficial temporal fat pad is clearly seen.

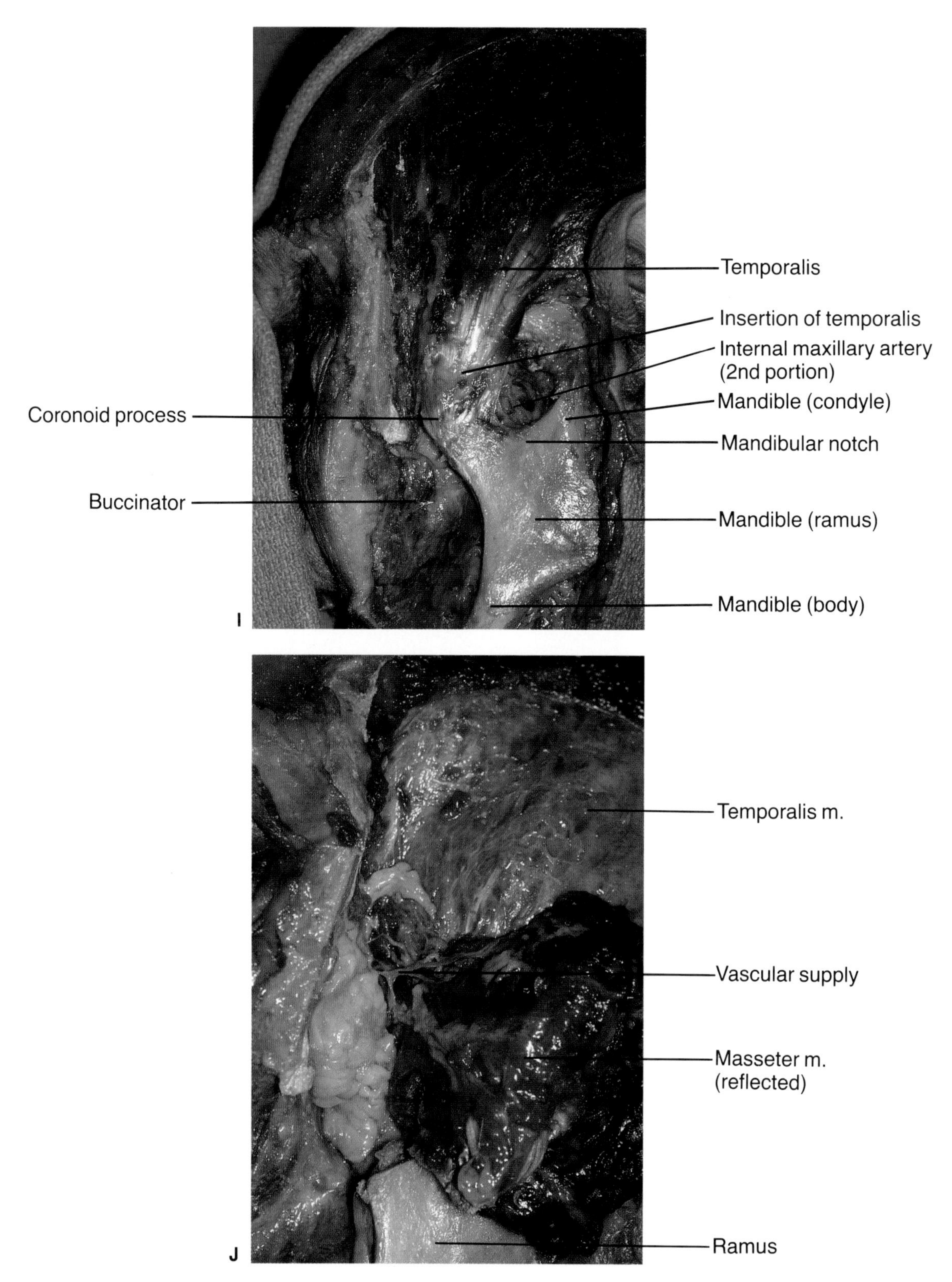

FIG. 6.15. *Continued.* **I:** The superficial temporal fat pad has been removed demonstrating the insertion of the temporalis into the coronoid process of the mandible. **J:** Part of the blood supply to the temporalis muscle is seen.

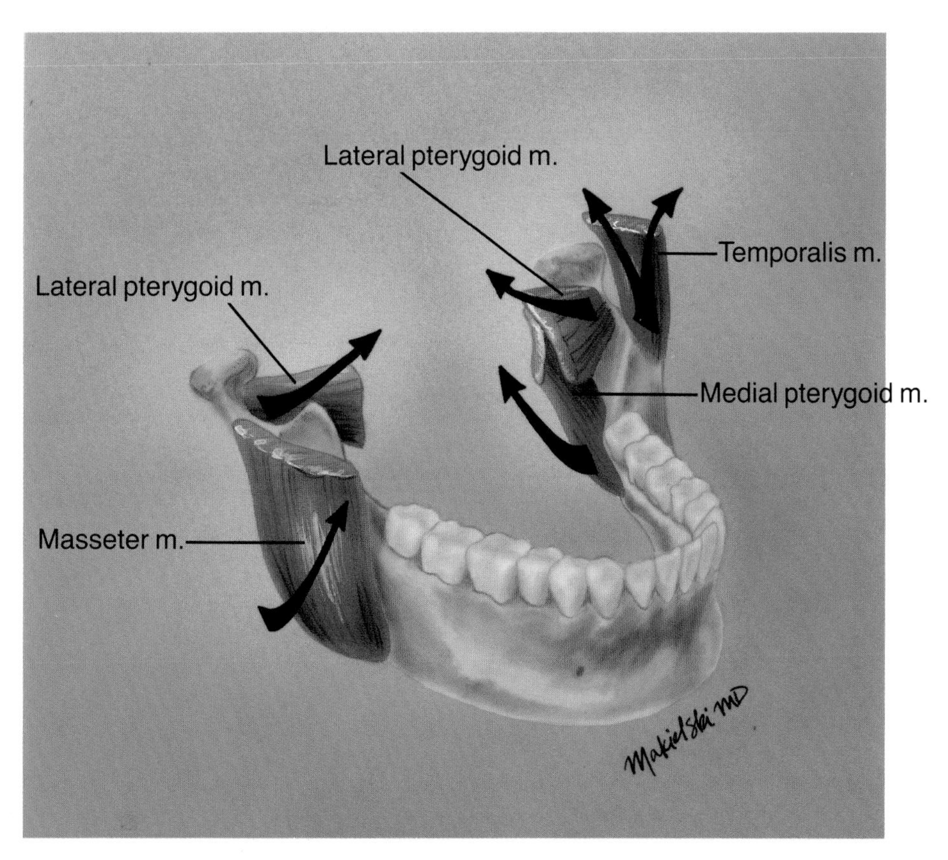

FIG. 6.16. Direction of muscle pull on the mandible: elevators. The masseter, the medial pterygoid, and the temporalis muscles elevate the mandible in normal function. The temporalis and the masseter muscles also function as mandibular retractors. The lateral pterygoid muscle is a protrusor. When the mandible has been fractured, all of these muscles function as elevators and tend to displace the fragments superiorly, medially, and anteriorly.

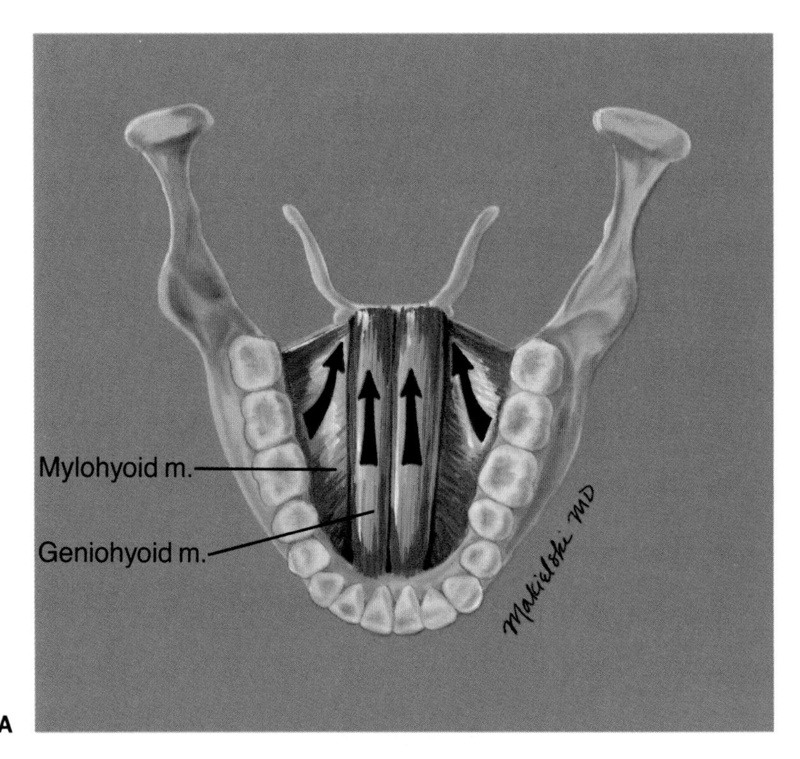

A

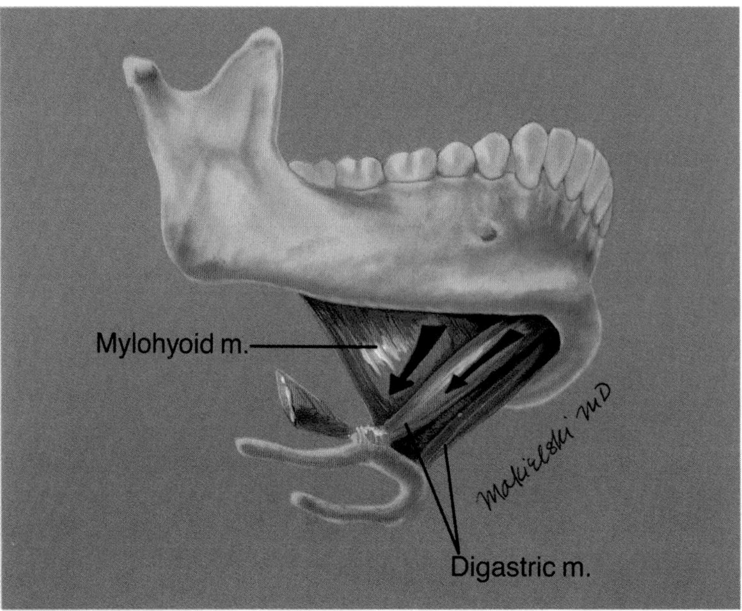

B

FIG. 6.17. Direction of muscle pull on the mandible: depressors. The anterior belly of the digastric, the geniohyoid, and, to a lesser degree, the mylohyoid muscles function as mandibular depressors and retractors. In the fractured mandible, these muscles pull the fragments downward, medially, and posteriorly. **A:** Superior view. **B:** Inferior view.

CHAPTER 7

Facial Nerve

The pathways of the facial nerve from the stylomastoid foramen to the undersurface of the muscles it innervates are quite variable. A knowledge of the landmarks for the common branching patterns and the depth of the nerve in specific regions is essential for safe and effective face-lift, facial reanimation, trauma repair, and other facial plastic surgery.

The main trunk of the nerve exits the stylomastoid foramen and immediately enters the parotid gland. The most consistent and reliable landmark for identification of the facial nerve is the tympanomastoid suture (Fig. 7.1). This suture line between the tympanic and the mastoid portions of the temporal bone points to the stylomastoid foramen, which is 6 to 8 mm beneath (medial to) the "drop-off point" of the tympanomastoid suture (Fig. 7.2) (Tabb and Tannehill, 1973). Other landmarks are helpful but less reliable. The main trunk can generally be found about midway between the cartilaginous pointer of the external auditory canal and the posterior belly of the digastric, where it attaches to the mastoid tip. The styloid process is deep to the main trunk.

BRANCHES OF THE FACIAL NERVE

The five commonly listed major branches of the facial nerve are the temporal (or frontal), the zygomatic, the buccal, (marginal) mandibular, and the cervical rami. In reality, the main trunk usually divides within the parotid into superior (temporofacial) and inferior (cervicofacial) divisions, but the branching patterns then become quite variable. Davis et al. (1956) classified these patterns into six types; although of interest, this type of classification is not of much practical benefit. A cadaver dissection (Fig. 7.3) shows a typical configuration. There are frequent anastomoses between the buccal and the zygomatic branches. The cervical branches are relatively unimportant. The temporal and the mandibular branches are most at risk of injury and are usually terminal branches without anastomotic connections. In thin patients the lower buccal and the upper mandibular branches can be damaged during surgery where they cross over the masseter muscle and the buccal fat pad (Fig. 7.4).

Temporal Branches

The temporal branches of the facial nerve have perhaps the most complex anatomy both in terms of horizontal branching patterns and in their relationship to fascial and muscular layers. The literature is somewhat contradictory, particularly concerning surface landmarks to determine the course of the nerve, but some useful guidelines are available (Fig. 7.5).

The superficial to deep relationships of the temporal branches have been described in Chapter 5. To summarize: the nerve rami exit the parotid, run within the SMAS over the zygomatic arch and the temple area, and enter the undersurface of the frontalis muscle. The nerve branches are superficial to both layers of the deep temporalis fascia. Thus, to avoid injury to the temporalis branch of the nerve when elevating flaps, one should undermine either in the immediate subcutaneous plane or deep to the SMAS on the temporalis fascia. Particular care is required over the zygomatic arch. These relationships are demonstrated in the dissections seen in Figure 7.6.

Mandibular Branch

Familiarity with the location of the marginal mandibular nerve is essential when operating in the lower face (Fig. 7.7). If the marginal mandibular nerve is injured during surgery, the resulting paralysis of the muscles that depress the corner of the mouth is quite deforming. Dingman and Grabb (1962) noted in a large cadaver study that, posterior to the facial artery, the marginal mandibular nerve passed above the inferior border of the mandible in 81% of dissections. Anterior to the facial artery, all of the mandibular nerve branches that innervated the mouth depressors passed above the lower border of the mandible. The only nerve branches that pass below the mandible anterior to the facial artery innervated the platysma and were therefore not of major surgical concern. The nerve was superficial to the posterior facial vein in 98% of the cases and superficial to the anterior facial vein in 100% of the cases.

The same study showed that the mandibular nerve may have one (21%), two (67%), three (9%), or four (3%) major branches. In 5% of the cases there was anastomosis between the buccal and the mandibular rami.

Baker and Conley (1979) write that, in their clinical experience, the mandibular branch of the facial nerve is usually 1 to 2 cm below the lower border of the mandible and can be as much as 3 or 4 cm below it. The mandibular branch is deep to the platysma muscle and therefore fairly well protected throughout its course along the mandible (Fig. 7.8). As it approaches the mouth it becomes more superficial and enters the undersurface of the depressor muscles (Fig. 7.9). Liebman et al. (1988) performed serial sections of cadavers and described the depth of the marginal nerve around the mouth (Fig. 7.7).

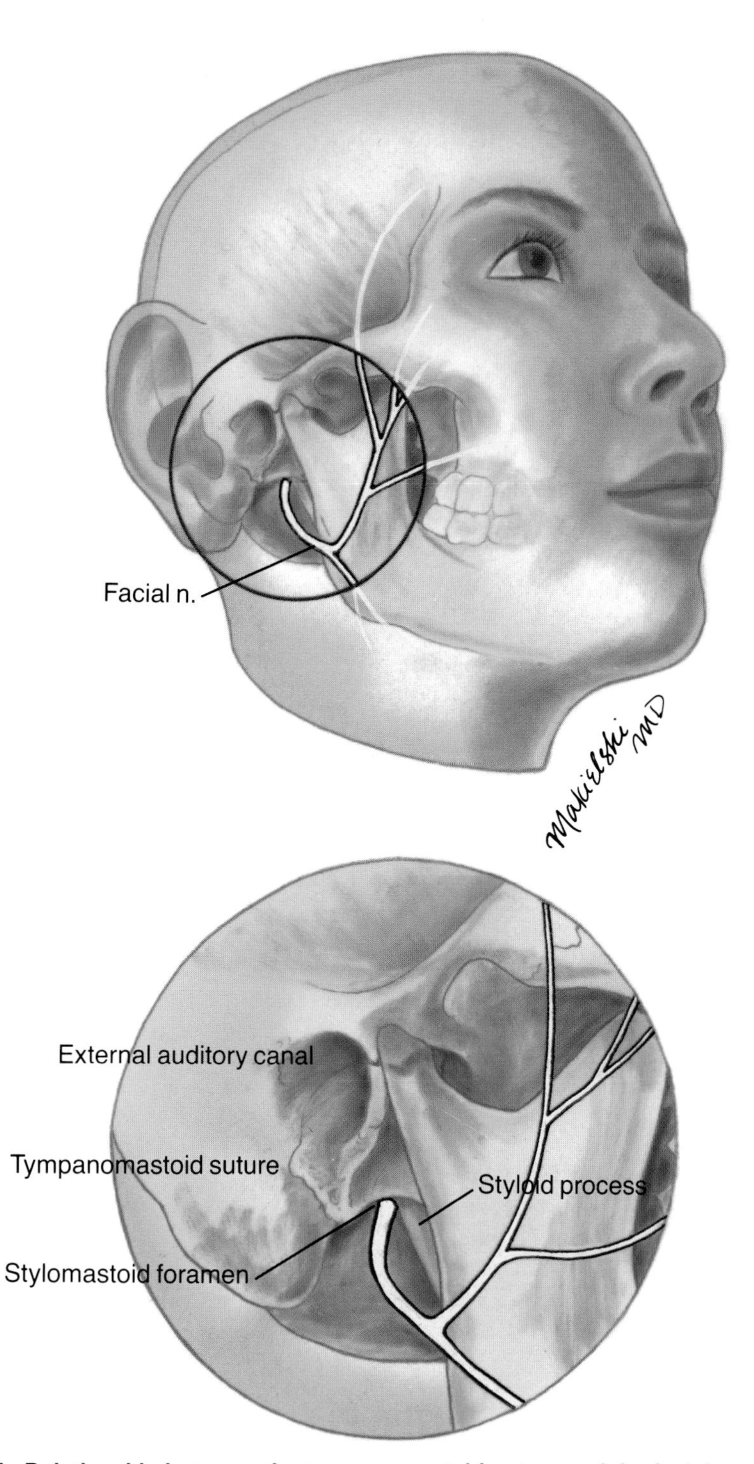

FIG. 7.1. Relationship between the tympanomastoid suture and the facial nerve. The tympanomastoid suture is the most reliable landmark to the facial nerve as it exits the stylomastoid foramen. The cartilaginous "pointer" of the ear is a helpful but less constant guide. The styloid process is deep to the main nerve trunk.

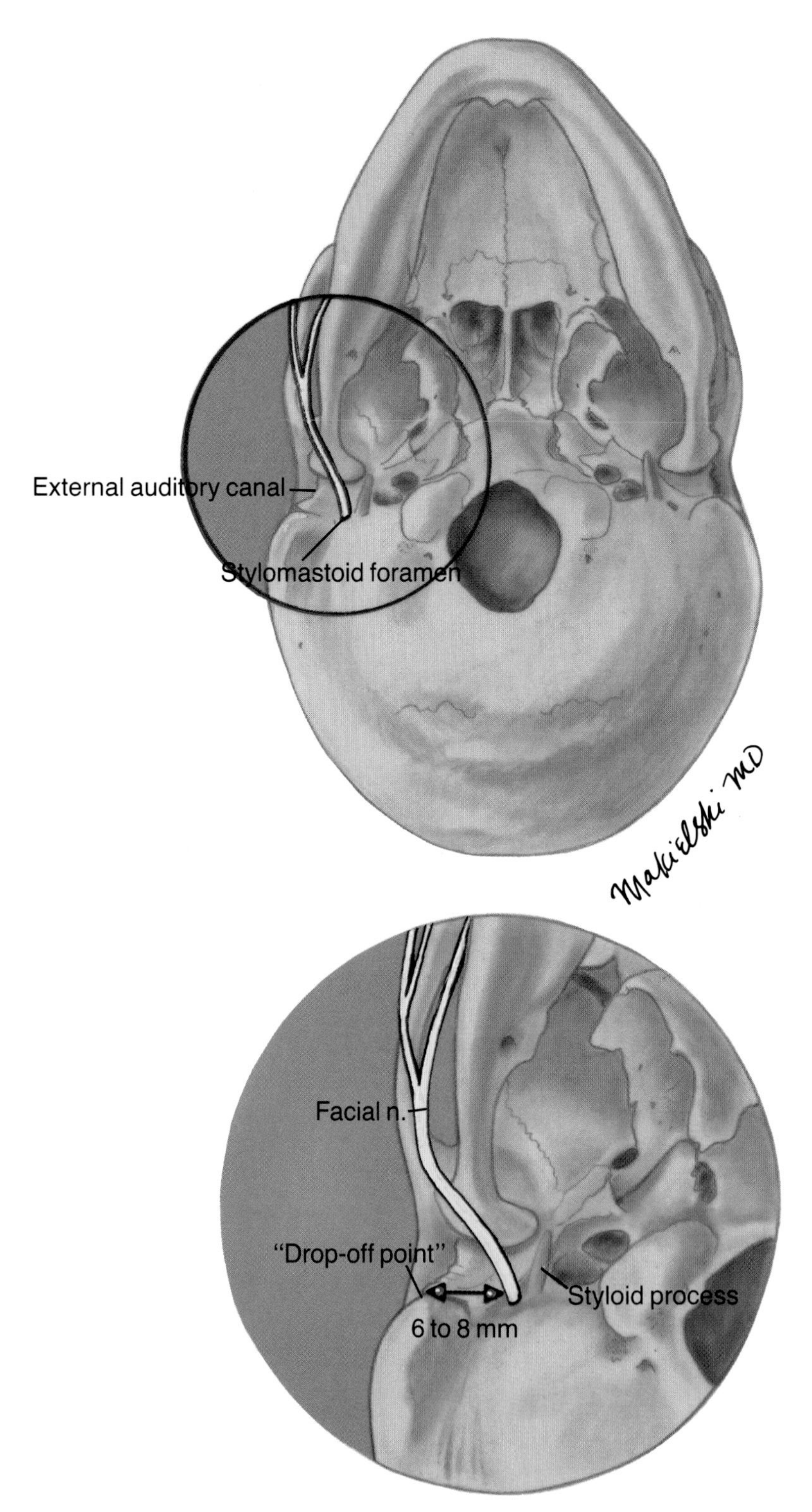

FIG. 7.2. Stylomastoid foramen: base view. The stylomastoid foramen and the facial nerve are 6 to 8 mm from the "drop-off point" of the tympanomastoid suture. Following the tympanomastoid fissure to the stylomastoid foramen will lead directly to the main trunk of the facial nerve, even in cases in which significant soft-tissue distortion is present.

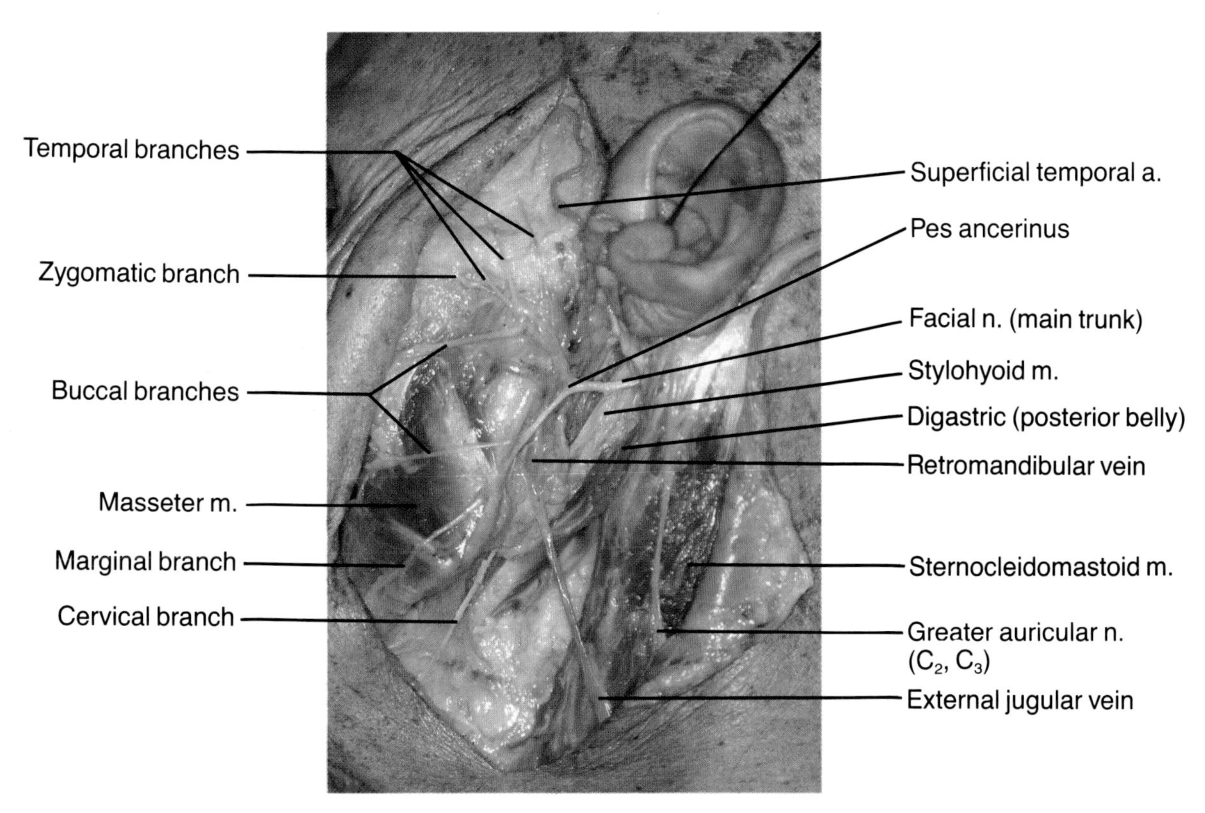

FIG. 7.3. Facial nerve after a total parotidectomy. The facial nerve exits the stylomastoid foramen and divides into inferior and superior segments at the pes ancerinus. In this case note the multiple temporal branches and the relationship of the nerve branches to the posterior belly of the digastric. The buccal branches overlie the masseter. The greater auricular nerve courses over the sternocleidomastoid muscle.

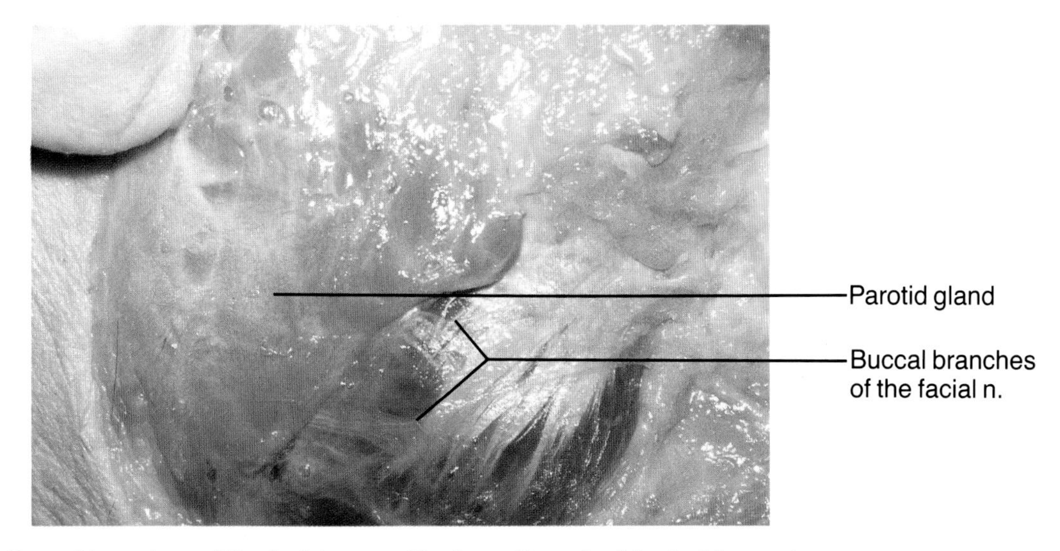

FIG. 7.4. Buccal branches of the facial nerve. The buccal branch of the facial nerve is seen exiting the parotid gland and coursing over the masseter muscle.

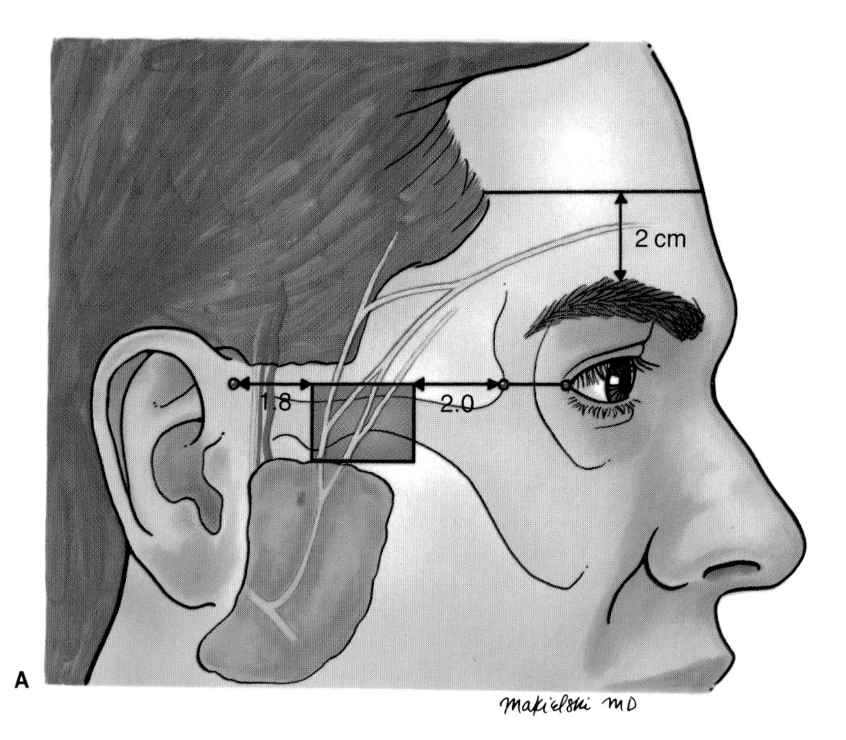

FIG. 7.5. Landmarks for the temporal branch of the facial nerve. A: The branching pattern in the temporal region was described by Bernstein and Nelson (1984) from a cadaver study. There are usually four rami that cross the zygomatic arch, although this varies from three to five. The most posterior ramus is always anterior to the superficial temporal vessels. A point on the anterior hairline at the level of the outer canthus consistently marks the junction of the posterior and the middle rami of the nerve. The most anterior ramus is on average 2 cm posterior to the anterior end of the zygomatic arch (it is sometimes described as running tangentially across the arch one fingerbreadth behind the lateral orbital rim). Another helpful, but not always accurate, guideline is that the nerve runs beneath a line 2 cm above the brow. At the temple, the nerve runs in the SMAS, inferior to the temporofrontal branches of the superficial temporal vessels. The anterior temporal hairline represents the lateral aspect of the frontalis muscle; medial to this, the nerve is deep to the muscle and is relatively protected. The danger area is seen in purple. Over the zygomatic arch the nerve is particularly vulnerable. It lies in the SMAS just beneath the subcutaneous fat and immediately overlying the bony prominence of the zygoma.

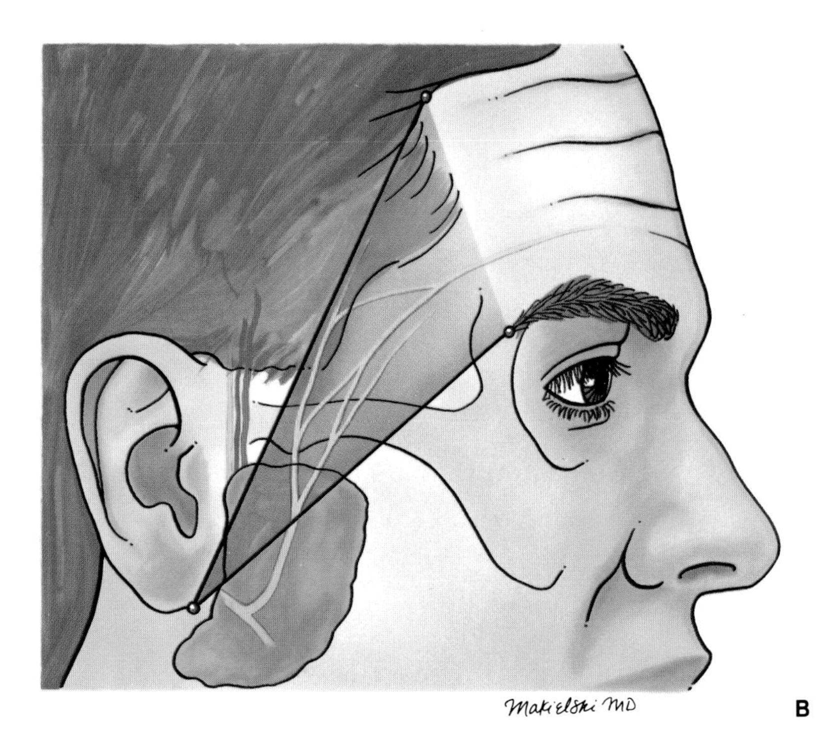

FIG. 7.5. *Continued.* **B:** Correia and Zani (1973) describe the temporal branches as lying between two diverging lines drawn from the earlobe to the lateral brow and to the lateral end of the highest forehead crease; although generally accurate, this rule ignores the auricular ramus. In the majority of this triangular area (seen in purple), the nerve is immediately subcutaneous and not protected by overlying muscle or parotid.

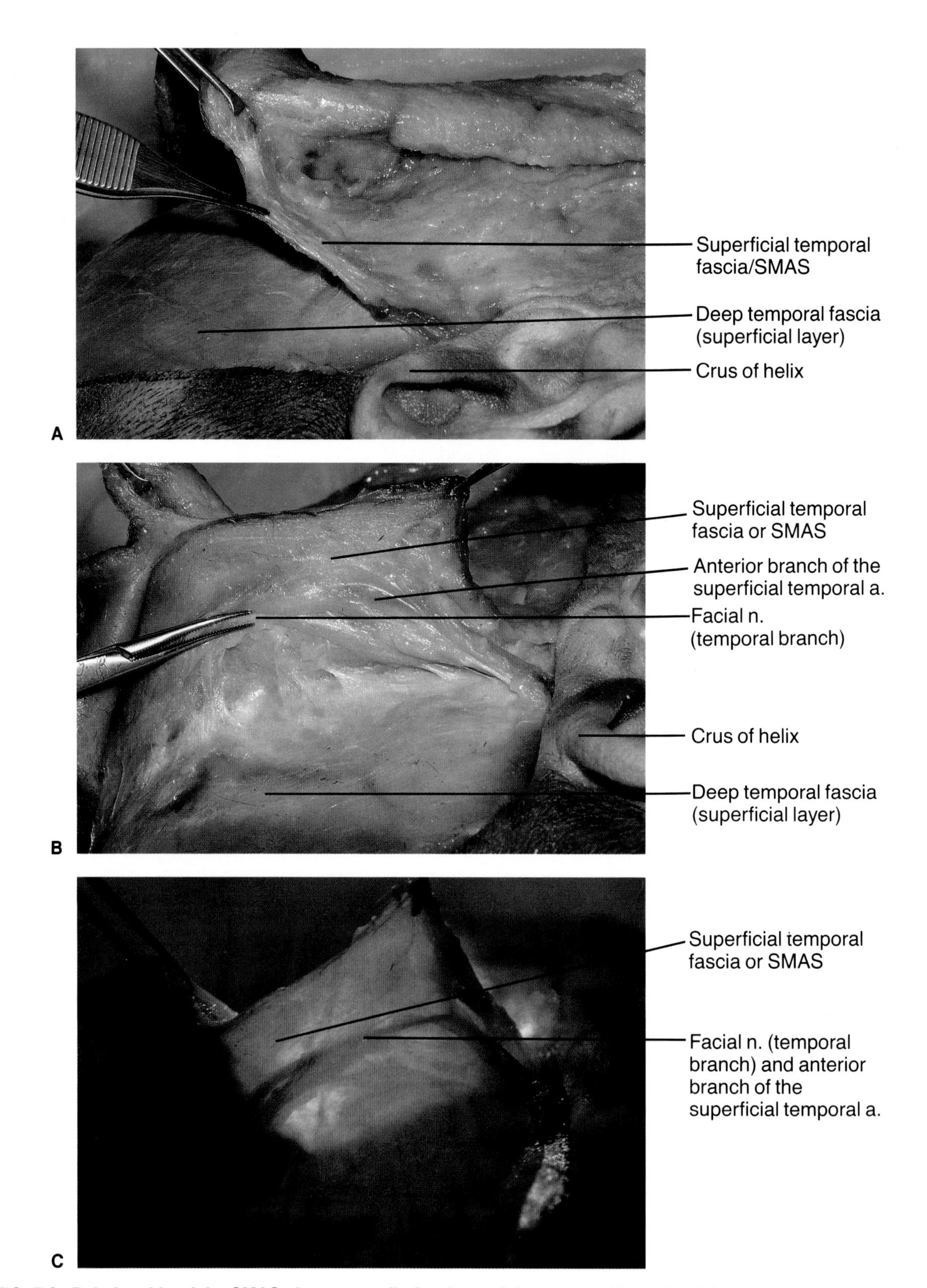

FIG. 7.6. Relationship of the SMAS, the temporalis fascia, and the temporal branch of the facial nerve. A: A bicoronal flap has been elevated in the subgaleal plane. Over the temporalis muscle, dissection is deep to the superficial temporal fascia, which carries the temporal branch of the facial nerve. A skin flap has been elevated inferiorly. The superficial temporal fascia, which is continuous with the SMAS inferiorly, is seen posteriorly (retracted with clamps). **B:** The same dissection is seen from above demonstrating the temporal branch of the facial nerve (hemostat) running in the SMAS with the anterior branch of the superficial temporal artery. **C:** The SMAS is transilluminated to demonstrate the temporal branch of the facial nerve and the accompanying artery.

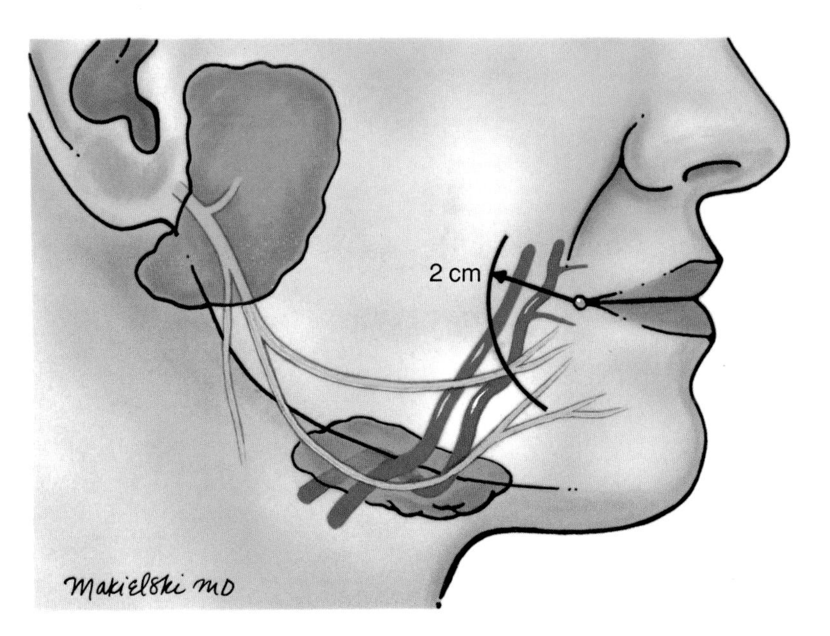

FIG. 7.7. Landmarks for the marginal mandibular branch of the facial nerve. The mandibular branch is located deep to the platysma until it reaches a point about 2 cm lateral to the corner of the mouth. At this point the nerve becomes more superficial and penetrates the undersurface of the facial mimetic muscles. Thus, in a face-lift or other skin-flap procedure, the mandibular nerve is fairly well protected if the dissection is subcutaneous and at least 2 cm lateral to the oral commissure (Liebman et al. 1988).

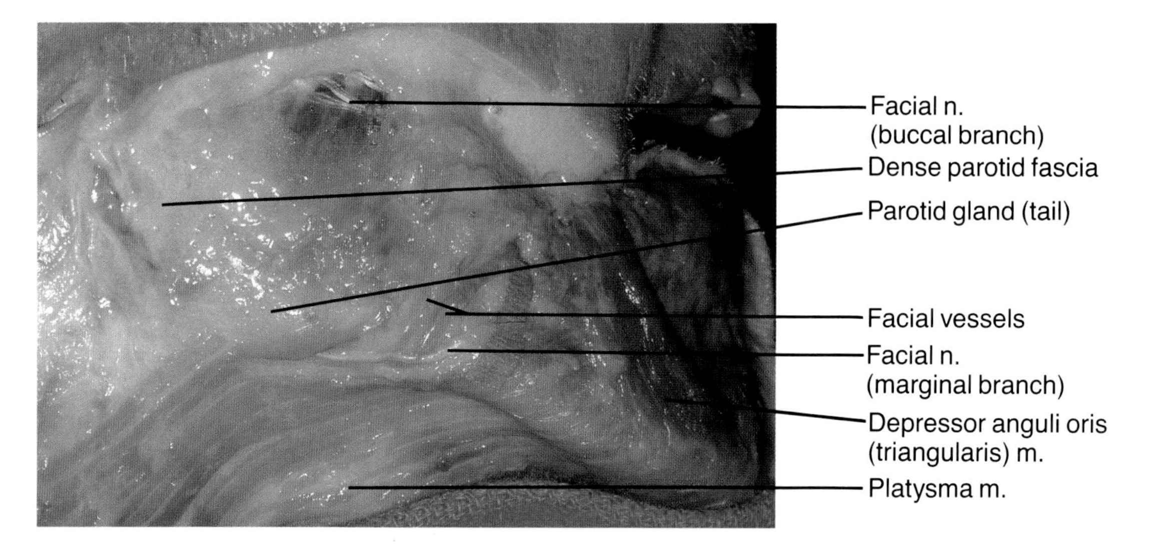

FIG. 7.8. Dissection of the marginal branch of the facial nerve showing the relationship of the platysma and the facial vessels. The marginal branch of the facial nerve is seen as it emerges along the anterior border of the parotid gland. The nerve crosses the inferior border of the mandible lateral to the facial vessels to supply the muscles of the lower lip and the chin.

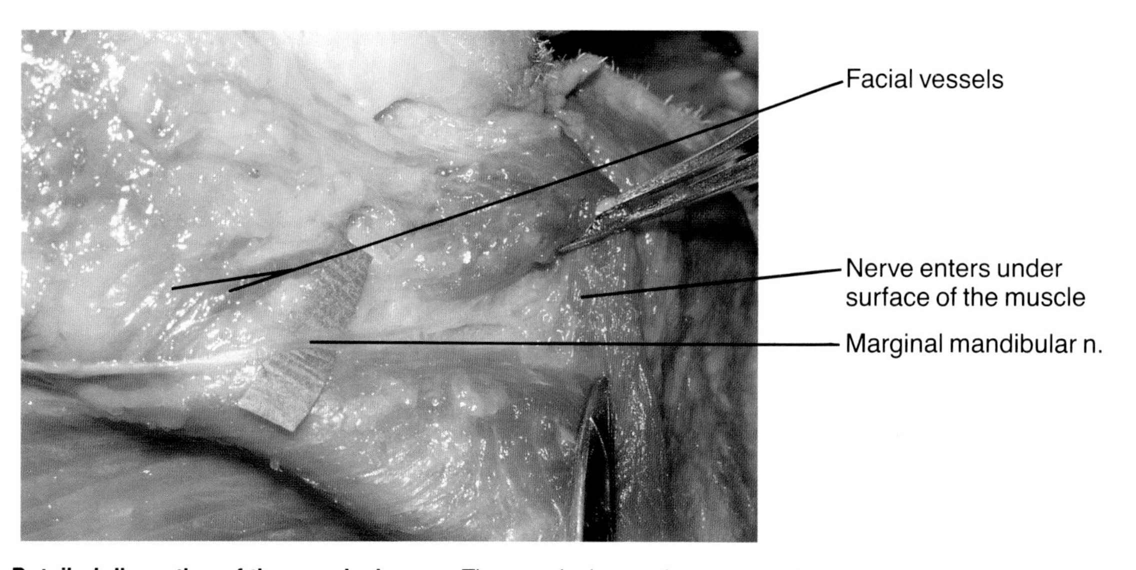

FIG. 7.9. Detailed dissection of the marginal nerve. The marginal nerve is seen entering the undersurface of the depressor muscles of the mouth.

CHAPTER 8

Facial Sensory Innervation

The sensation of the head and the neck can be generally divided into two groups: the face, and the scalp and the neck (Fig. 8.1). The sensory innervation of the face is from the trigeminal nerve (cranial nervc V). Each division of the trigeminal nerve supplies all tissues from the skin to the mucosa and corresponds approximately to the three embryological divisions of the face: the frontonasal process (V1), the maxillary process (V2), and the mandibular process (V3). The sensation of the posterior scalp, the lower border of the mandible, and the neck is supplied by branches of the cervical nerves and the plexus. The scalp and the neck dermatomes begin with the second cervical nerve (C2) at the top of the scalp (C1 does not have a cutaneous branch), extending down to C4 in the supraclavicular region (Fig. 8.2). Because the focus of this book is on the face, the cervical nerves will not be discussed further.

DIVISIONS OF THE TRIGEMINAL NERVE

The ophthalmic division of the trigeminal nerve (V1) supplies sensation to the forehead, the anterior scalp, and the nasal dorsum via the supratrochlear and the supraorbital nerves (Fig. 8.3). The infratrochlear and the dorsal nasal nerves and the terminal branches of the anterior ethmoid nerve provide additional sensation to the nasal dorsum. The upper nasal and the septal mucosa are innervated by the anterior ethmoid nerve. The supraorbital and the ethmoid nerves provide sensation to the frontal sinus, the ethmoid sinuses, and the sphenoid sinus. The ophthalmic nerve also provides sensation to the skin and the conjunctiva of the upper eyelid and the ciliary nerve to the cornea and the eyeball. A branch goes to the dura of the tentorium cerebelli.

The maxillary division of the trigeminal nerve (V2) innervates the cheek and the side of the face, the conjunctiva and the skin of the lower eyelid, the side of the nose and the nasal vestibule, and the mucosa and the skin of the upper lip via the zygomaticotemporal and the zygomaticofacial nerves and the infraorbital nerve (Fig. 8.4). Three branches of the infraorbital nerve are of note: the anterior superior, the middle superior, and the posterior superior alveolar nerves. The anterior superior alveolar nerve supplies the maxillary incisors and canines, as well as mucosa of the anterior maxillary sinus, the nasal cavity, and the gingiva. The middle superior alveolar nerve supplies the maxillary premolars, the first molar, and some of the maxillary sinus; there is substantial variability in the scope of innervation of this nerve. The posterior superior alveolar nerve innervates the maxillary molar teeth, the maxillary molar buccal gingiva, the buccal mucosa, and much of the maxillary sinus mucosa.

The pterygopalatine (sphenopalatine) nerve provides sensation to the mucoperiosteum of the nasal cavity, the septum, the palate, the sphenoid and the ethmoid sinuses, and the nasopharynx. Specific branches include the nasopalatine nerve to the anterior palate and the septum, the greater palatine nerve (anterior palatine nerve) to the posterior hard palate, and the lesser palatine nerve (middle and posterior palatine nerves) to the soft palate and the tonsillar mucosa.

The mandibular branch of the trigeminal nerve (V3) has both a sensory and a motor component. There are four clinically important sensory branches to the mandibular nerve: the inferior alveolar nerve, the lingual nerve, the buccal nerve, and the auriculotemporal nerve (Fig. 8.5). Areas supplied by the inferior alveolar nerve include the lower teeth, the gingiva, and the mandible. One of its branches, the mental nerve, supplies sensation to the mucosa and the skin of the lower lip and the chin. Innervation to the anterior two-thirds of the tongue, the floor of the mouth, and the lingual gingiva is provided by the lingual nerve. The buccal nerve supplies the buccal mucosa and the cheek skin. The auriculotemporal nerve provides sensation for the temporal skin and portions of the external auditory canal and the tympanic membrane. The auriculotemporal nerve also carries some motor fibers to the parotid gland and thus can cause Frey's syndrome when the fibers are misdirected to the subcutaneous sweat glands. The motor branch of the mandibular nerve supplies the four muscles of mastication: the temporalis, the masseter, and the medial and the lateral pterygoids. It also innervates the two tensors, the tensor tympani and the tensor veli palatini. The mylohyoid muscle and the anterior belly of the diagastric also receive their motor innervation from the mandibular nerve.

SPECIAL AREAS

Nose

The soft tissue over the external nose has a complex innervation that does not lend itself well to blocks of specific nerves. The nasal dorsum, as previously noted, is supplied by the ophthalmic nerve through its infratrochlear branch to the nasal root, and through the terminal branch of the anterior ethmoid as it emerges between the nasal bones and the upper lateral cartilage to supply the more inferior nasal dorsum. The lateral nasal area and most of the tip are supplied by the infraorbital nerve. The columella is often difficult to block and might derive part of its innervation from the terminal nasal branch of the anterior superior alveolar nerve, which supplies the mucous membrane of the anterior floor of the nose and the nasal spine (Fig. 8.6).

Oral Cavity

The oral cavity has a diverse and complex innervation (Fig. 8.7). The lips and the anterior oral mucosa can be mostly anesthetized with a block of the infraorbital and the mental nerves. These two nerves (along with the supraorbital nerve) lie in a sagittal plane approximately at the midpupillary line (Fig. 3.1). The tongue is primarily innervated by the lingual nerve and the teeth by the alveolar nerve. The palate is largely innervated by the nasopalatines and the greater and the lesser palatine nerves.

Maxillary Sinus

The posterior superior alveolar nerve supplies sensation to most of the maxillary sinus. Several other nerves contribute to sensation of the antrum (Fig. 8.8). These contributions are important to remember when doing surgical procedures on the maxillary sinus under local anesthesia. The anterior superior alveolar nerve innervates the anterior portion of the maxillary sinus, and the middle superior alveolar nerve contributes to innervation of some of the mucosa. The maxillary ostium is supplied by the greater palatine nerve. The infundibulum, however, receives its innervation from the anterior ethmoid nerve, a branch of V1.

Ethmoid Sinuses

The ethmoid sinuses are innervated by two branches of the ophthalmic nerve, the anterior ethmoid and the supratrochlear, as well as a branch of the maxillary nerve, the sphenopalatine.

Frontal Sinus

All of the sensation of the frontal sinus is provided by branches of the ophthalmic nerve, the anterior ethmoid, and the supraorbital, with the latter also supplying the nasofrontal duct.

Sphenoid Sinus

Both the maxillary and the ophthalmic divisions innervate the sphenoid sinus, specifically the sphenopalatine, the posterior ethmoid, and the supraorbital nerves.

Eyelids

The lower lid is supplied by the infraorbital nerve. The upper lid receives twigs superomedially from the supraorbital, the supratrochlear, and the infratrochlear nerves, and laterally from the lacrimal nerve.

Forehead

Clinically the supraorbital and the supratrochlear nerves are most commonly encountered when performing a coronal forehead flap for reconstructive or cosmetic purposes (Fig. 8.9).

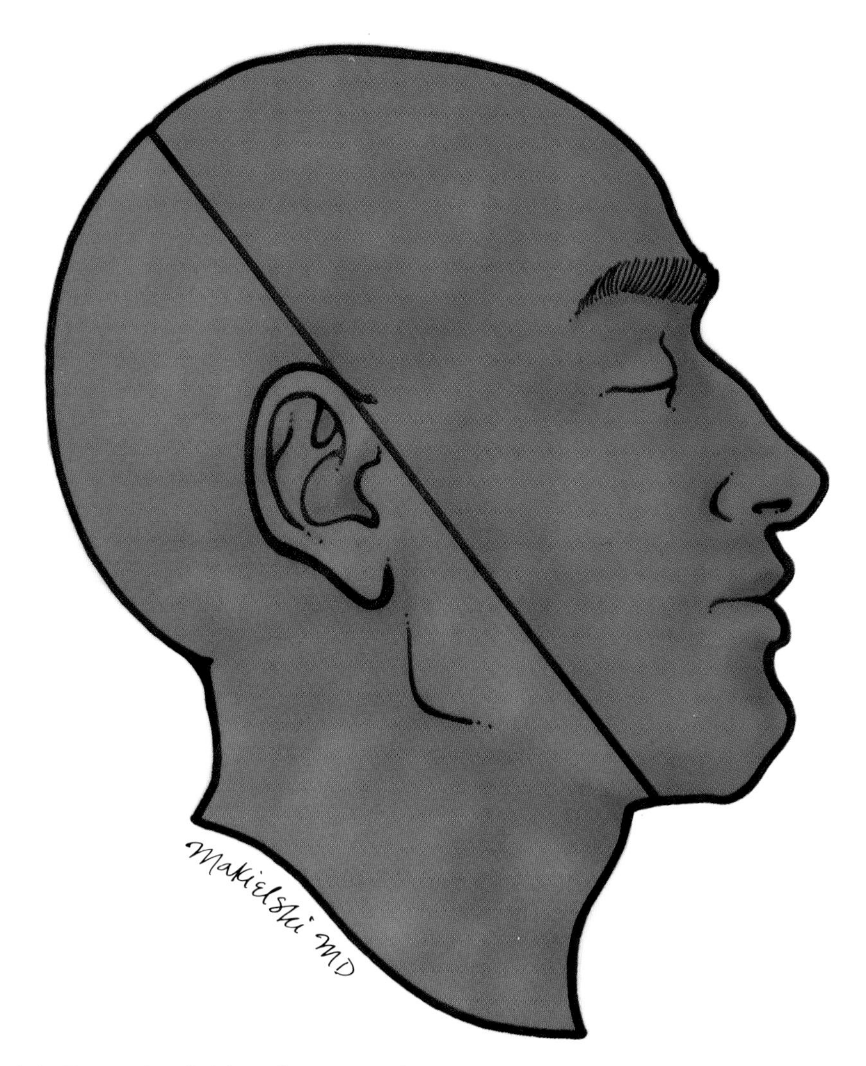

FIG. 8.1. Two major divisions for sensory innervation of the head and the neck. The trigeminal nerve through its three major divisions supplies sensation to the majority of the face and the anterior scalp. The cervical branches supply the neck and the posterior scalp.

Purple = Trigeminal n.
Red = Cervical n.

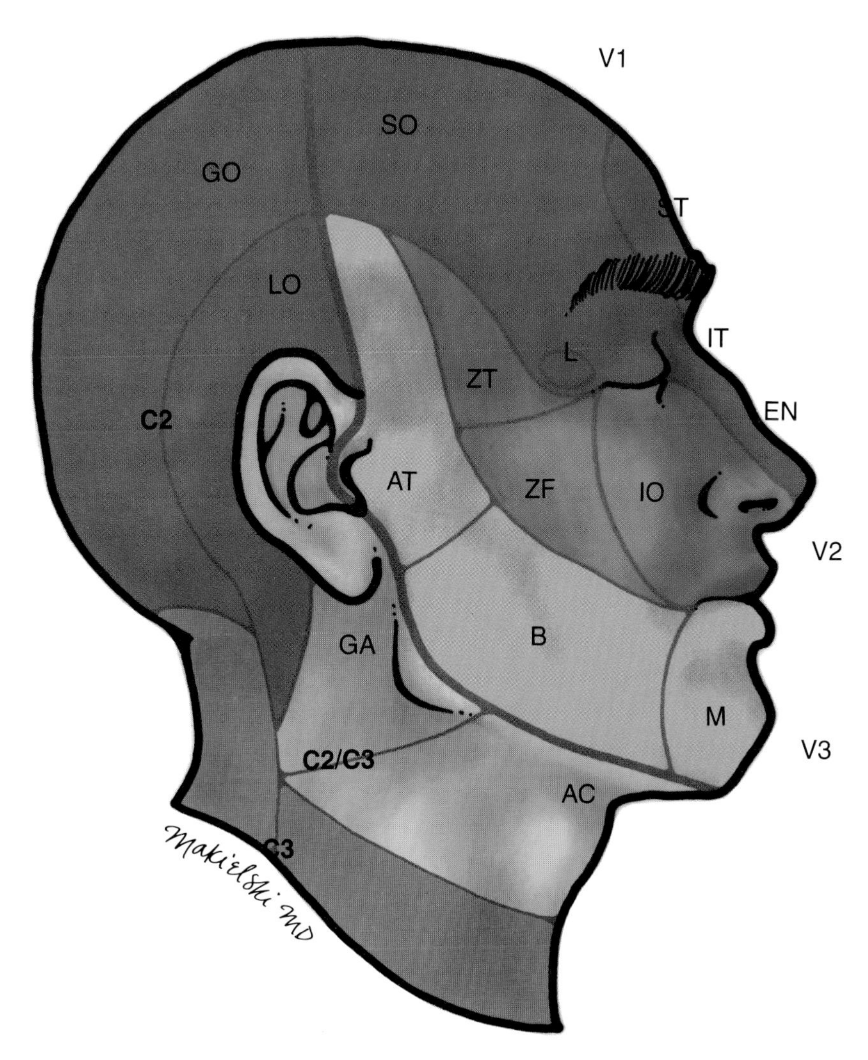

FIG. 8.2. Dermatomes of the head and the neck.

Trigeminal nerve:
- V1 (ophthalmic n.) = Purple
- Supraorbital n. (SO)
- Supratrochlear n. (ST)
- Infratrochlear n. (IT)
- External nasal n. (EN)
- Lacrimal n. (L)

- V2 (maxillary n.) = Blue
- Zygomaticotemporal n. (ZT)
- Zygomaticofacial n. (ZF)
- Infraorbital n. (IO)

- V3 (mandibular n.) = Green
- Auriculotemporal n. (AT)
- Buccal n. (B)
- Mental n. (M)

Cervical nerves:

C2 = Red
- Greater occipital n. (GO)
- Lesser occipital n. (LO)

C2/C3 = Yellow
- Greater auricular n. (GA)
- Anterior cutaneous n. (AC)

C3 = Orange

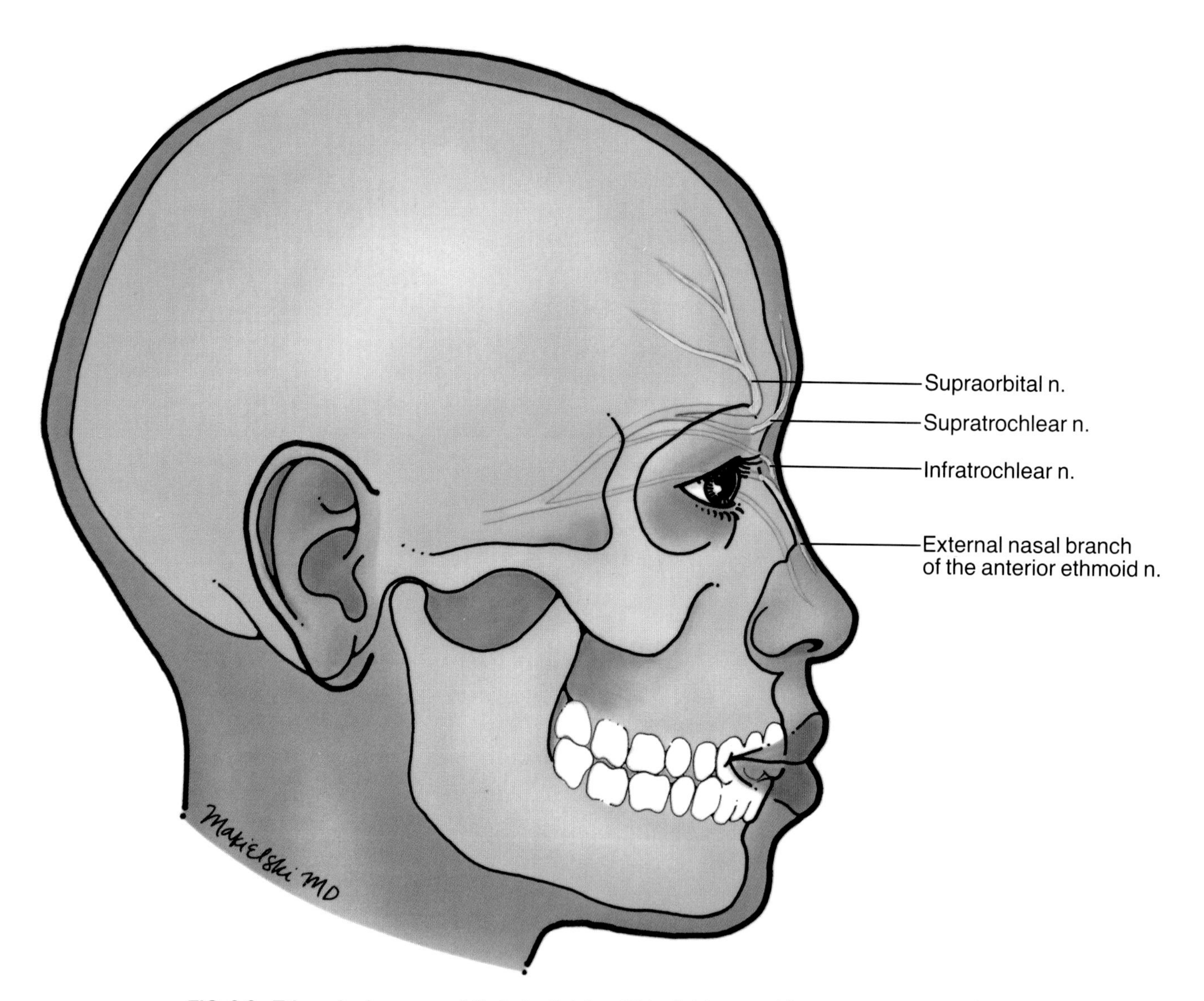

FIG. 8.3. Trigeminal nerve: ophthalmic division. This division provides cutaneous sensation to the forehead, the anterior scalp, and the nasal dorsum through the supraorbital, the supratrochlear, the infratrochlear, and the external nasal branches of the anterior ethmoid nerves. A small lacrimal branch at the lateral canthus also provides sensation to the upper lid.

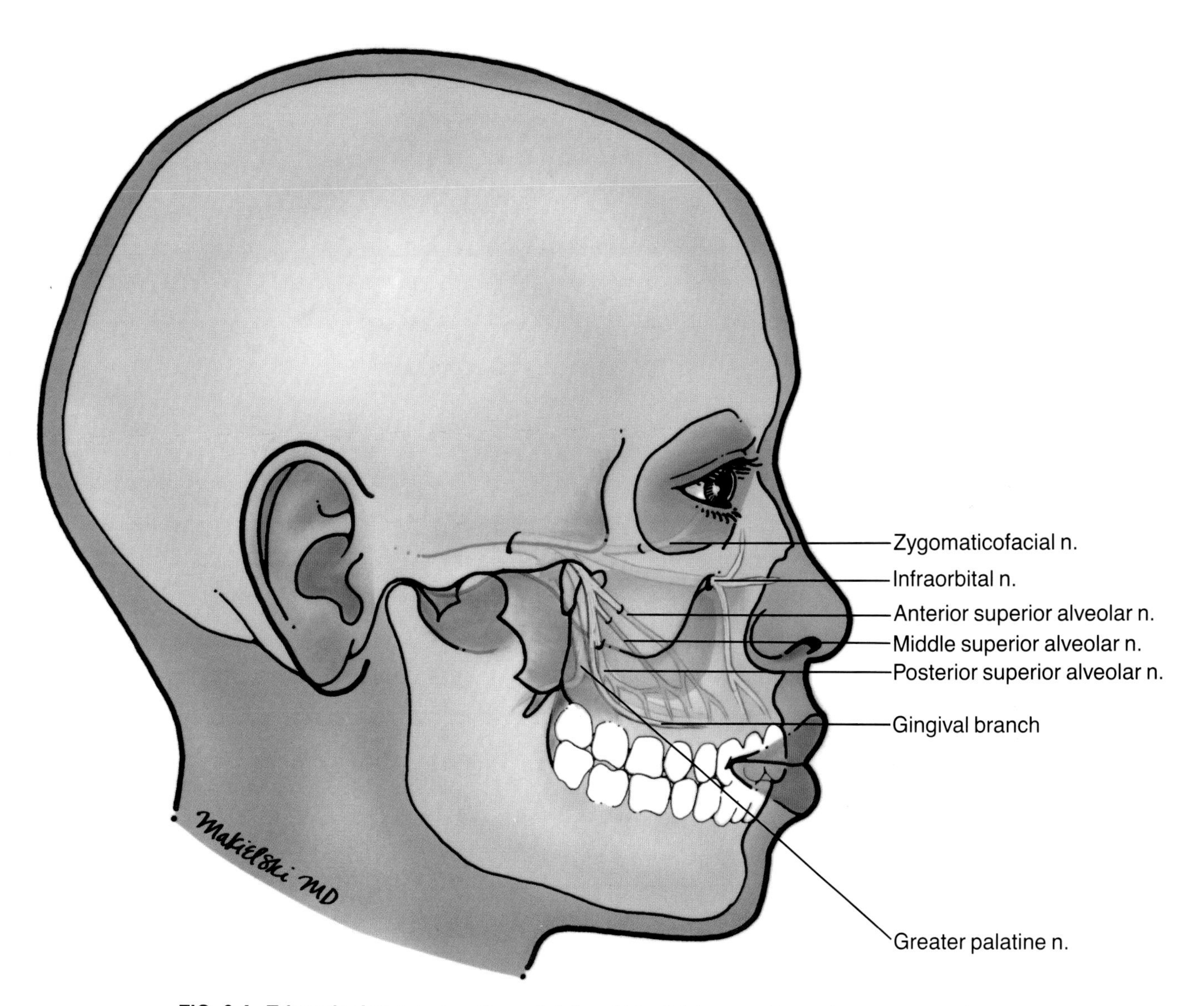

FIG. 8.4. Trigeminal nerve: maxillary division. This division provides cutaneous sensation to the cheek, the side of the face, the lower eyelid, the side of nose, and the upper lip. There is significant variation in the exact distribution of the anterior superior alveolar, the middle superior alveolar, and the posterior superior alveolar nerves, which supply sensation to the maxillary dentition.

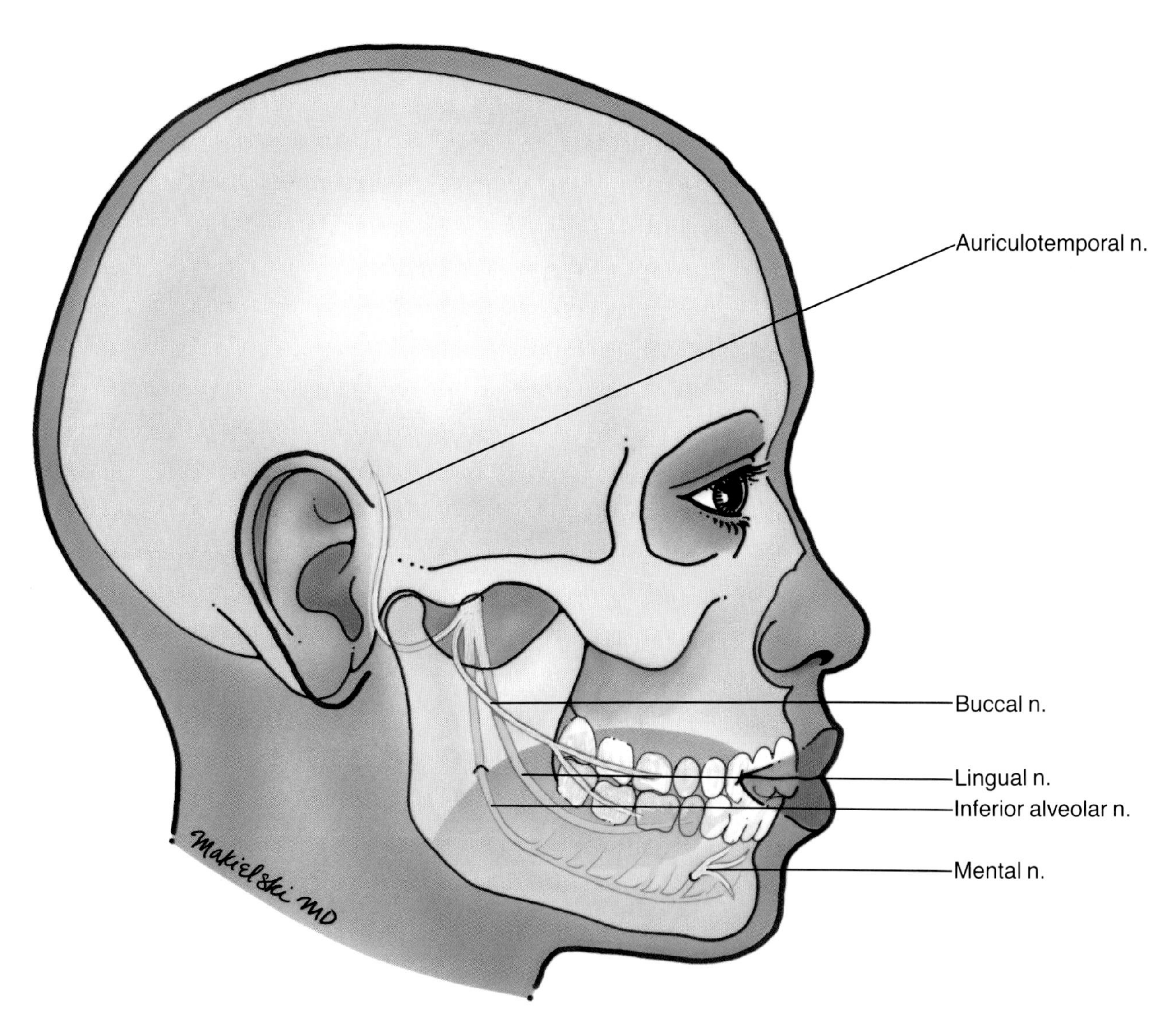

FIG. 8.5. Trigeminal nerve: mandibular division. This division supplies sensation to the lower lip, the cheek, the temple, the anterior two-thirds of the tongue, the floor of mouth, and the mandibular dentition.

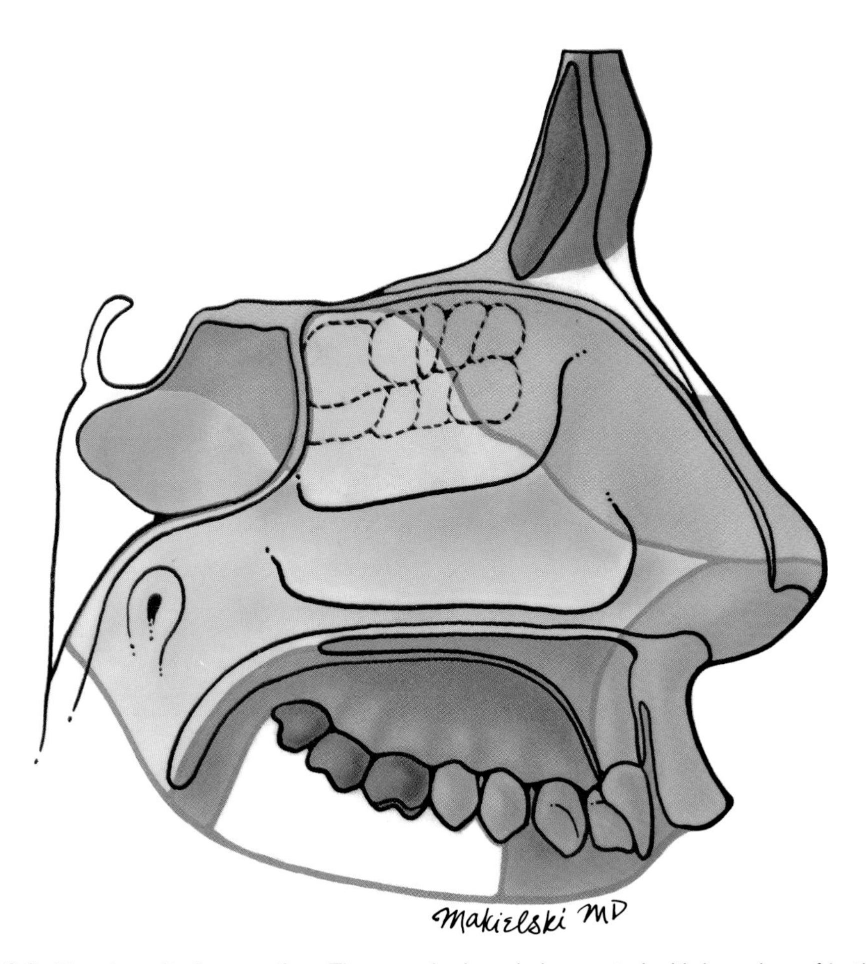

FIG. 8.6. Nasal cavity innervation. The nose is densely innervated with branches of both V1 and V2. The anterior ethmoid nerve (V1) is seen in green; the sphenopalatine nerve (V2) is seen in yellow; the infraorbital, the anterior, the middle, and the superior alveolar nerves are seen in blue; the posterior superior alveolar nerve is seen in purple. To anesthetize the nasal mucosa topically, anesthesia is applied along the roof of the nose to block the anterior ethmoid nerve, posterior to the middle turbinate to block the sphenopalatine branches, and along the floor to block the remaining branches of V2.

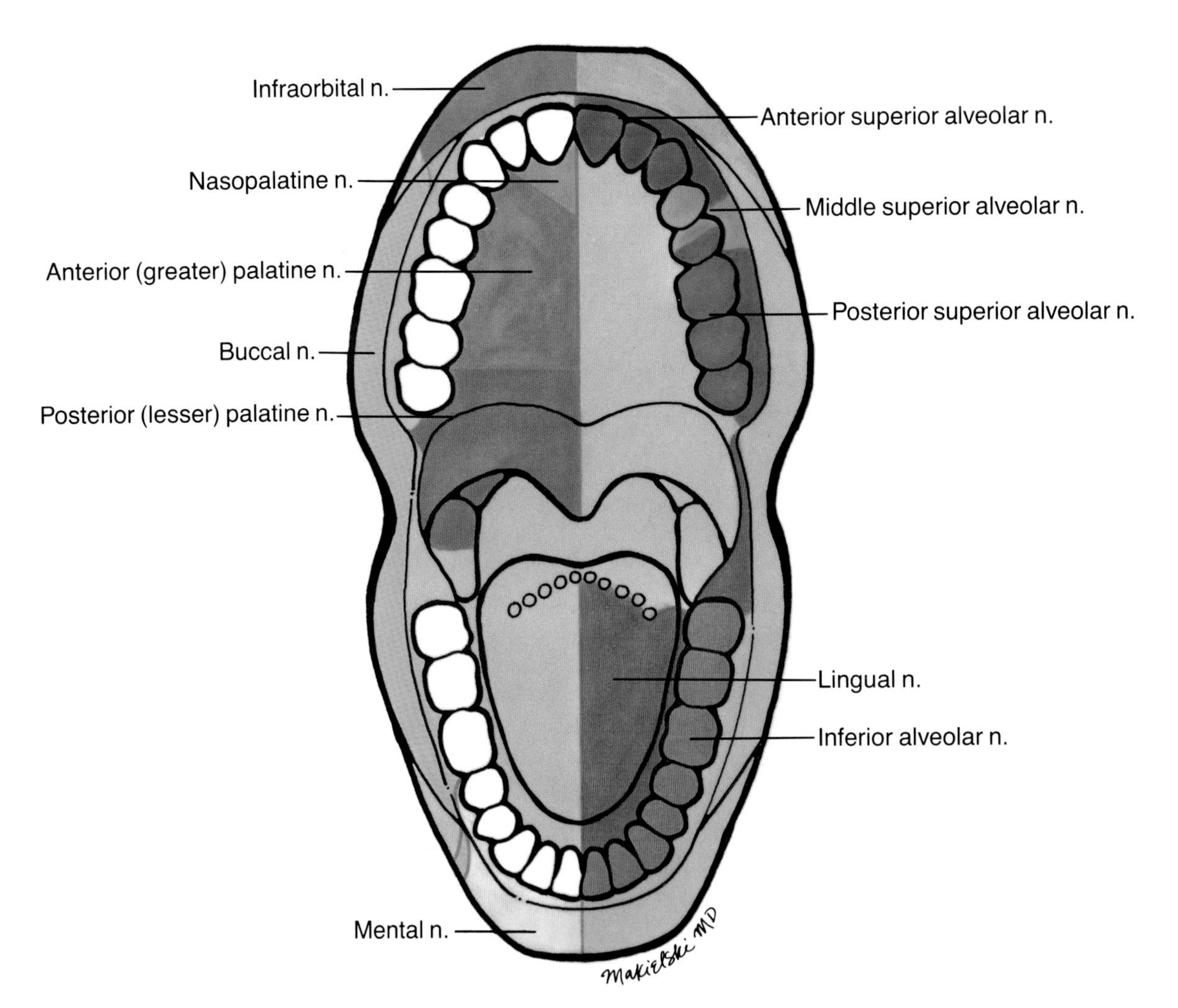

FIG. 8.7. Oral cavity innervation.

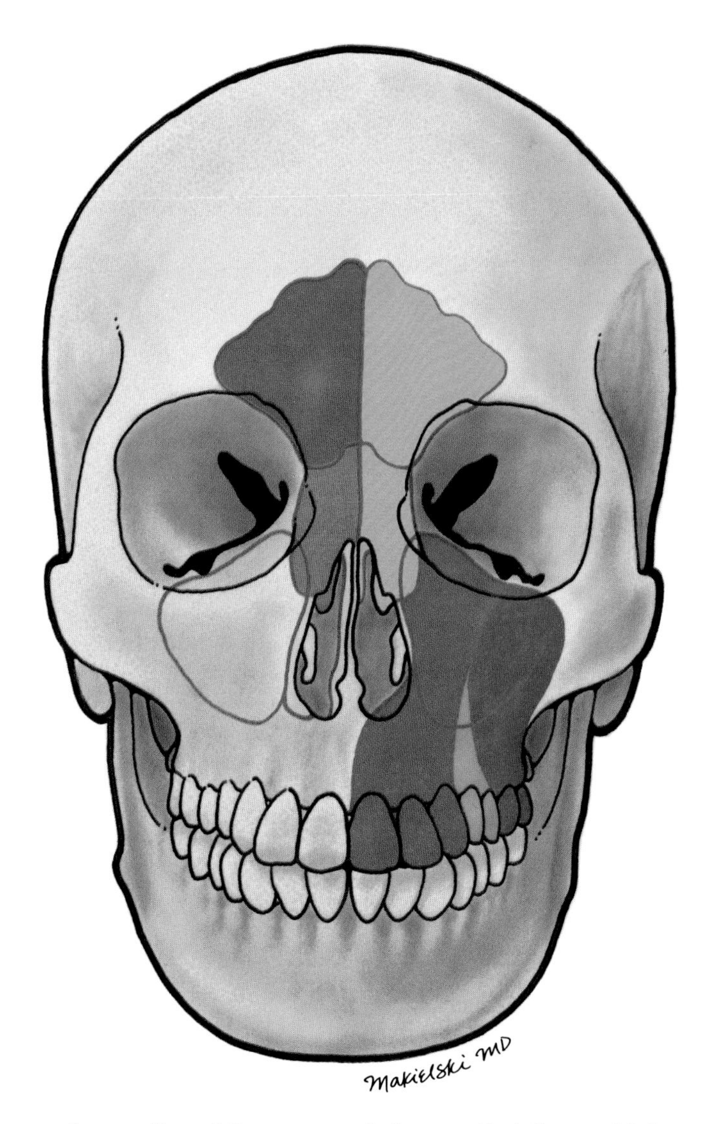

FIG. 8.8. Sensory innervation of the paranasal sinuses. Red, Supraorbital n.; orange, Overlap of supraorbital n. and sphenopalatine n.; yellow, Sphenopalatine n.; green, Anterior ethmoid n.; purple, Anterior superior alveolar n.; blue, Middle superior alveolar n.; navy, Posterior superior alveolar n.

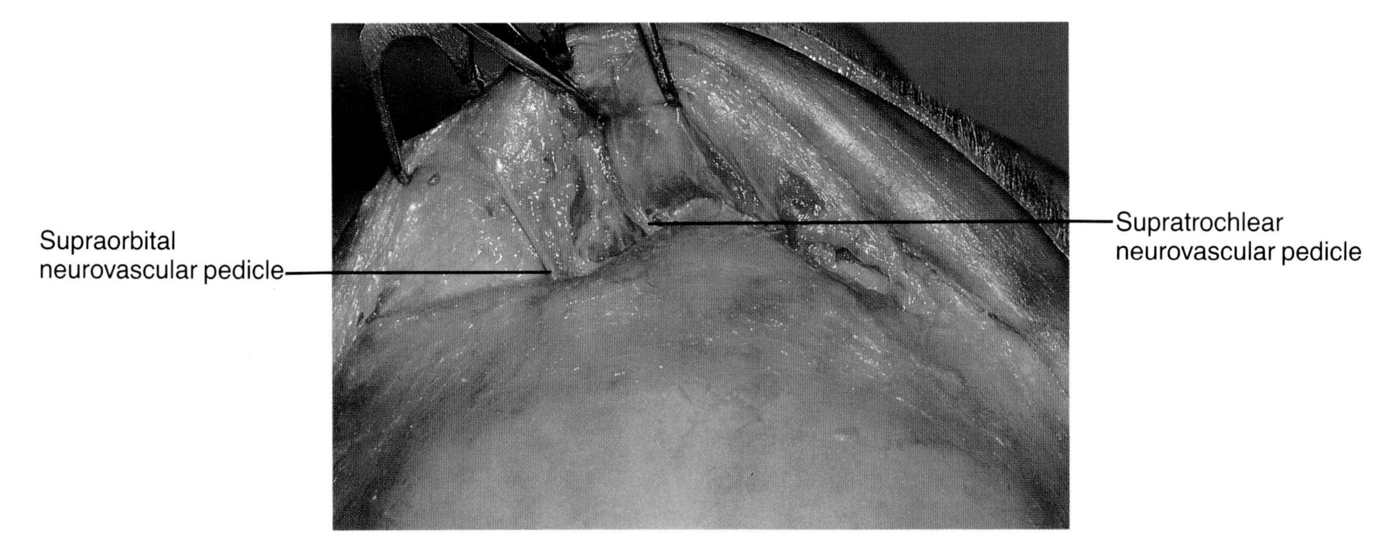

FIG. 8.9. Supraorbital and supratrochlear neurovascular pedicles. From the coronal approach, one can easily identify and protect the neurovascular pedicles. The supraorbital nerve and the vessels originate in a foramen or a notch at approximately the midpupillary line. By using blunt dissection over the supraorbital rim (e.g., with a cotton tip applicator), the pedicle can be identified and mobilized as necessary for reconstructive orbital procedures.

CHAPTER 9

Vascular Patterns of the Face

The vascular system is a three-dimensional linkage of neighboring vascular territories. Major distributing vessels travel close to the underlying bony skeleton, emerging from the deep tissues at fixed points, either from bony foramina or where the deep fascia approximates the underlying bone. For instance, several branches of the external carotid artery (the occipital, the posterior auricular, and the superficial temporal arteries) emerge near the skull base, whereas the facial artery emerges along the lower border of the mandible. Other important facial vessels arise at the bony foramina of the mandible and the maxilla (the infraorbital and the mental arteries) and at the orbital margins (the supraorbital, the supratrochlear, the dorsal nasal, and the medial palpebral arteries).

The distributing arteries then branch into smaller perforator vessels that comprise the basis for the muscular and the cutaneous vascular systems. Typically, perforator branches enter their respective muscle groups on the deep surface, closely following the connective tissue architecture of the muscle as they divide and subdivide into smaller nutrient vessels before extending to the cutaneous surface.

ANGIOSOMES

Of importance in understanding the vascular system is the concept of the angiosome, introduced by Taylor and Palmer (1987). An angiosome is a composite segment of bone, muscle, nerve, and overlying skin, supplied by a common source vessel. Angiosomes of the face, as represented by the cutaneous vascular territories, are shown in Figure 9.1. These segments form the theoretic basis for the design of complex tissue flaps. A flap that is based on the source vessel of a single angiosome can frequently incorporate some tissues from an adjacent angiosome; nutrient blood will be supplied to the adjacent angiosome through collateral channels.

THE CUTANEOUS CIRCULATION

The anatomy of the cutaneous vascular system is of particular importance in the design of successful tissue flaps.

Arteries

Branches of perforator arteries extend to the cutaneous surface as either septocutaneous or musculocutaneous arteries (Fig. 9.2). Septocutaneous vessels travel generally within the fascia (septa) of the muscle. In contrast, musculocutaneous vessels pass directly through the muscle tissue, providing multiple nutrient branches to the surrounding muscle as they travel vertically to the cutaneous tissues.

Blood vessels serving the cutaneous tissues have traditionally been classified into two separate groups: the direct and the indirect cutaneous arteries. Direct cutaneous arteries are vessels whose primary purpose is to provide blood flow to the skin. These arteries often are septocutaneous branches from deep-lying source arteries; they may also arise from perforating musculocutaneous branches or as a continuation of a source vessel such as the superficial temporal artery. Direct cutaneous vessels frequently emerge near points of muscular attachment to surrounding bone and fascia.

Indirect arteries form a secondary echelon of vessels that supply the cutaneous vascular plexus. These vessels are frequently terminal branches of musculocutaneous vessels that supply large immobile muscles; they are of little use in the design of artery-based skin flaps.

Veins

The venous system is of equal importance to the arterial system in the proper design and execution of various tissue flaps. Venous outflow from the skin occurs by way of a subdermal venous plexus that generally enters the deep venous system through valves in the venae communicantes. Venous drainage may also occur by way of the paired venae comitantes that accompany the small cutaneous arteries.

Tissue flaps in the face should be designed so that flow occurs in the proper direction through the valves in the venae communicantes. Flaps constructed in the scalp are rarely subject to problems with venous outflow because scalp veins do not possess valves.

Cutaneous Vascular Plexus

The cutaneous tissues and their nutrient vessels form a stack of interconnected vascular tissue planes called the vascular plexus, consisting of the fascial plexus, the subcutaneous plexus, and the subdermal plexus (Fig. 9.2). The deepest structure in the cutaneous vascular plexus is the fascial plexus, at the level of the deep muscle fascia. The fascial plexus derives its blood supply from small vessels that branch off septocutaneous and musculocutaneous arteries as they penetrate the muscular fascia, and from retrograde flow from the overlying subcutaneous plexus. By including the fascial plexus in the design of the fasciocutaneous flap, tissue survival may be improved (Tolhurst et al., 1983). Fascial flaps can serve as a highly vascularized bed for skin graft placement and have been liberally employed in reconstructive ear surgery for many years.

Overlying the fascial plexus is the subcutaneous plexus, a significant network of vessels corresponding to the level of the superficial fascia or SMAS. The vascular density of the subcutaneous plexus demonstrates tremendous regional variation and has a reciprocal relationship with the density of the overlying subdermal plexus.

The subdermal, or cutaneous, plexus is the most significant of the horizontal vascular layers and has a primary role in the distribution of blood to other regions of the cutaneous system. The subdermal plexus lies at the junction between the reticular dermis and the underlying subcutaneous fat. Clinically, this level corresponds with the phenomenon of dermal bleeding, which is often seen at the leading edge of skin flaps.

Superficial to these major subcutaneous plexus lie the dermal plexus and the closely associated subepidermal plexus. These layers perform two primary functions: the dermal plexus provides thermoregulation, whereas the capillary beds of the subepidermal plexus provide nutrients to the skin. The capillary density of the skin and the subcutaneous tissues is only a small fraction of that found in the muscular system. Furthermore, only a small percentage of blood flow is used for nutrient value; the remainder is used to regulate temperature and blood pressure.

REGIONAL VASCULAR ANATOMY

The majority of skin and subcutaneous tissue in the face is supplied by branches of the external carotid system (Fig. 9.3). The exception is a "mask-like" region of the central face that encompasses the eyes, the upper two-thirds of the nose, and the central forehead (Fig. 9.1). The ophthalmic branch of the internal carotid system provides the primary blood supply to this region, with anastomoses to the facial and the superficial temporal branches of the external carotid artery providing additional blood supply (Figs. 9.4 and 9.5).

The following anatomic regions have vascular systems of particular significance for flap design and other surgical procedures.

Temporal Area

The superficial temporal artery is the terminal branch of the external carotid system, arising deep within the parotid gland and emerging between the condyle of the mandible and the external auditory meatus. As the artery nears the zygomatic arch, it lies just deep to the dermis within the substance of the temporal-parietal fascia or SMAS; this fascia also transports the temporal branch of the facial nerve.

Prior to crossing the zygomatic arch, the superficial temporal artery generates a small horizontal branch known as the transverse facial artery (Fig. 9.6). The transverse facial vessel runs parallel and inferior to the zygomatic arch to supply the lateral canthus. At the superior border of the zygomatic arch, the superficial temporal artery gives off a second branch, the middle temporal artery. This artery immediately penetrates the superficial layer of the deep temporalis fascia to supply the superficial temporal fat pad and the deep temporalis fascia. It also contributes to perfusion of the temporalis muscle.

Within 2 cm of crossing the zygomatic arch, the superficial temporal artery divides into two or three terminal branches. If there are two branches (more common), the anterior branch supplies the forehead, forming anastomotic connections with the ipsilateral supraorbital and the supratrochlear vessels and the contralateral anterior branch of the superficial temporal artery. The posterior branch supplies a wide territory over the parietal skull and has rich anastomotic connections with adjacent and contralateral arteries. When a third branch exists, it runs just superior to the supraorbital ridge; it is called the horizontal branch. The size of this additional artery varies inversely with the ipsilateral supraorbital and supratrochlear vessels; its physiologic importance varies considerably.

Orbit and Eyelids

The vascular supply of the eyelids is derived from the medial and the lateral palpebral arteries at the medial and the lateral canthi. These supply the marginal arteries that lie superficial to the tarsal plates under the orbicularis oculi approximately 3 mm from the free edge of the lid. The upper lid also has a peripheral artery that runs transversely at the upper edge of the tarsal plate deep to the levator palpebrae superioris.

The medial palpebral artery is derived from the anastomosis of the infratrochlear and the angular arteries. The lateral palpebral artery is derived principally from the lacrimal artery. Other contributions to the vascular supply of the eyelids come from the supratrochlear and the supraorbital arteries superiorly and the infraorbital and the transverse facial arteries inferiorly.

The external vessels of the lids are primarily derived from the branches of the external carotid system, whereas those within the orbit are derived from the internal carotid. This region is therefore a potential area of anastomosis between the internal and the external carotid system. Since the central artery of the retina is a branch of the internal carotid, thrombi (from trauma, surgery, or other causes) may enter the central artery and produce blindness. This anastomosis is probably also the anatomic basis for the cases of blindness reported after the inadvertent intraarterial injection of various substances around the eye.

The veins of the eyelids drain into the angular and the ophthalmic veins medially and the superficial temporal vein laterally. The angular-ophthalmic anastomosis provides a valveless communication from the venous drainage of the skin of the medial canthal area to the cavernous sinus, creating a route for intracranial spread of orbital infection.

Nose

The dorsum of the nose is principally supplied by the lateral nasal branch of the facial artery, with contributions over the nasal root by the infratrochlear and the supratrochlear arteries. The terminal branch of the anterior ethmoid artery, the external nasal artery, emerges between the upper lateral cartilages and the nasal bones to supply a small area of the nasal dorsum. The anterior and the posterior ethmoid arteries supply the superior aspect of the lateral wall of the nose and are responsible for some cases of posterior epistaxis. The ethmoid arteries are branches of the internal carotid system, with anastomoses to the external carotid system in the nasal cavity. As a result, the potential for inadvertent intraarterial injection and embolization leading to blindness exists here as well as in the orbital region. The posteroinferior aspect of the lateral nasal wall is supplied by the sphenopalatine artery, which is also frequently implicated in posterior epistaxis. The columella is supplied by a twig from the septal branch of the superior labial artery. The venous drainage parallels the arterial supply.

Lips

The lips are supplied by the inferior and the superior labial arteries, which are branches of the facial artery. They run in or behind the fibers of the orbicularis oris muscle and thus are superficial relative to the labial mucosa (see Fig. 17.3). The venous drainage corresponds to the arterial supply.

VASCULAR ANATOMY AND FLAP DESIGN

The vascular supply to the face is extensive and contains several direct cutaneous arteries that are of value in the design of arterial skin flaps. The supraorbital and the supratrochlear arteries can support cutaneous flaps of the midline forehead. The frontal branch of the superficial temporal artery is important in the design of the forehead and the bipedicle visor flap. The parietal or the posterior branch of the superficial temporal artery may support various forms of hair-bearing tissue transfers.

Flaps can be divided into five categories, based on their vascular supply: random cutaneous, arterial cutaneous, fasciocutaneous, myocutaneous, or composite.

Random cutaneous flaps are supplied by perforating septocutaneous and musculocutaneous vessels entering at the anatomic base of the flap. Perfusion of the distal flap segment occurs by way of the cutaneous vascular plexus. Random flaps are used extensively on the face and encompass the majority of transposition, advancement, and rotation flaps performed in this region (Fig. 9.7).

Arterial cutaneous flaps are based on the presence of axially aligned direct cutaneous arteries that permit large areas of skin to be raised. The flap may extend beyond the termination of the artery, depending on the degree of collateral flow through the cutaneous vascular plexus. Arterial cutaneous flaps may be used as pedicle flaps or as "free flaps" for microvascular transfer. Clinical examples include the midline forehead flap and the parascapular skin flap.

Fasciocutaneous flaps are designed to include the underlying muscular fascia. Flap survival is improved due to circulation provided by the fascial plexus and the adjacent subfascial course of arteries. Fasciocutaneous flaps are widely used in the scalp and in the extremities, where direct cutaneous vessels hug the fascial layer for a distance prior to entering the subcutaneous tissue. Fasciocutaneous flaps may be used as pedicled flaps or in microvascular transfer. Clinical examples of the fasciocutaneous flap include the free radial forearm flap and the deltopectoral flap.

Myocutaneous flaps, which include the underlying muscle and fascia, are based on an essential vascular pedicle. The skin is supplied by perforating musculocutaneous vessels. Skin survival is limited to the approximate dimensions of viable muscle incorporated into the flap. Few true myocutaneous flaps can be developed in the face since the facial skin is not ordinarily supplied by perforating vessels from the underlying musculature. The nasalis flap is one myocutaneous flap that can be developed in this region. Myocutaneous flaps from other regions of the body can provide skin for mucosal lining and body resurfacing and provide muscle bulk for dead space closure and wound healing. Current trends favor using only part of a muscle, based on the patterns of vessels and nerves in the muscle. A useful classification of the vascular supply to muscles has been proposed by Mathes and Nahai (1981).

Composite flaps consisting of bone, muscle, and cutaneous tissue may be rotated on a vascular pedicle or may be transferred to distant sites utilizing microvascular anastomosis.

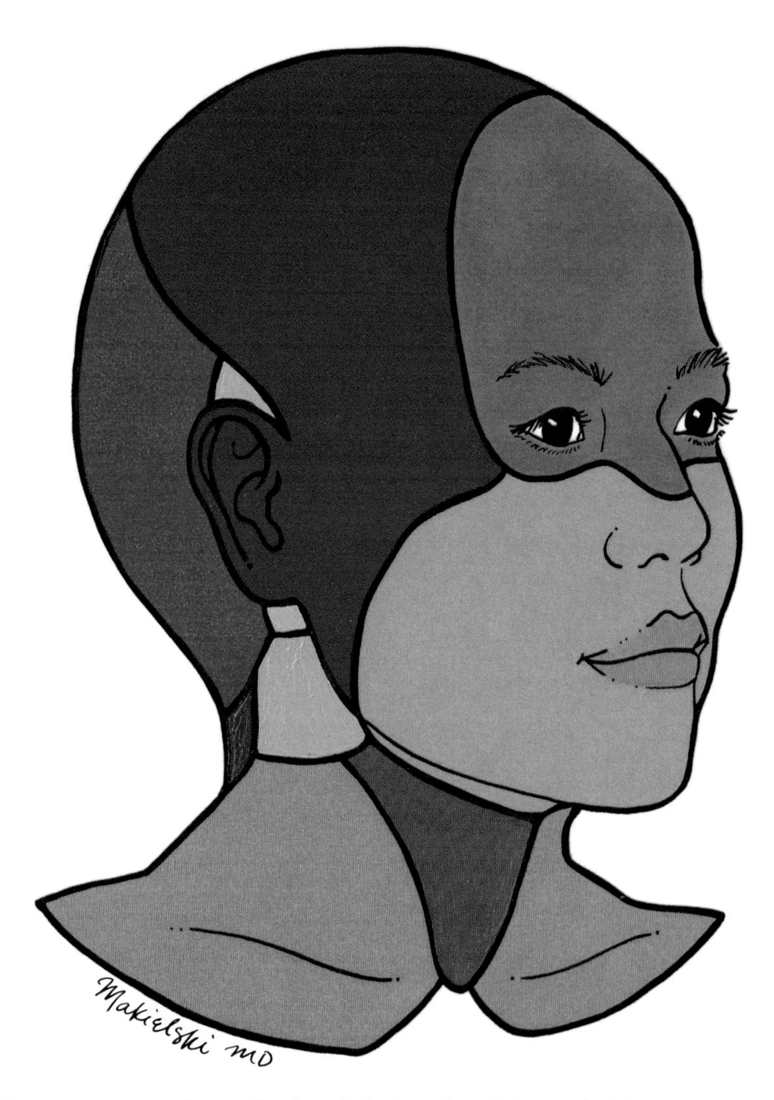

FIG. 9.1. Cutaneous vascular territories of the head and the neck. The majority of skin and the subcutaneous tissue is supplied by branches of the external carotid system. The exception to this rule lies within a "mask-like" region of the face, which includes the eyes, the upper nose, and the central forehead. This central region serves as an anastomotic link between the internal (ophthalmic artery) and external (facial, superficial temporal arteries) carotid systems.

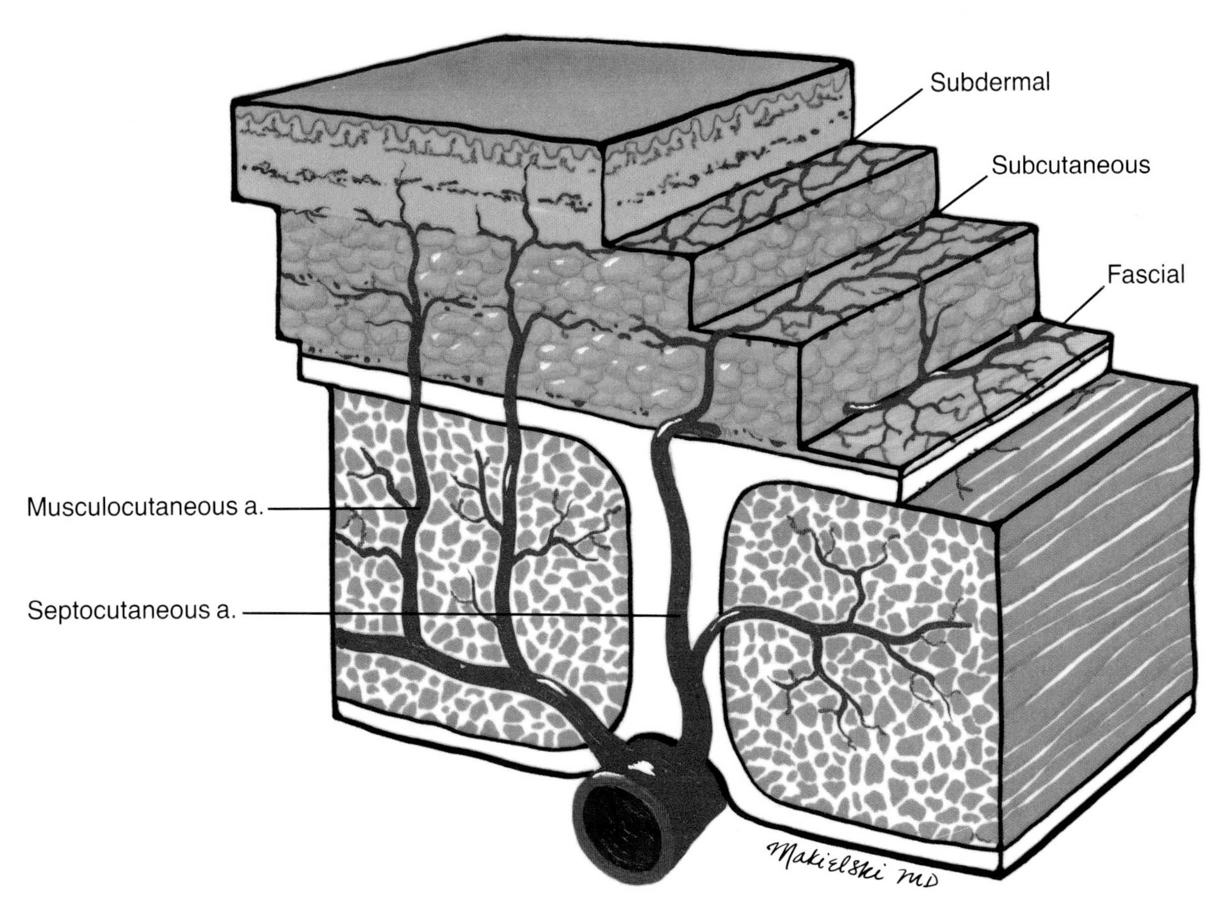

FIG. 9.2. Cutaneous vascular plexus. The cutaneous vascular plexus forms a stacked series of interconnected vascular tissue planes that derive their blood supply from septocutaneous and musculocutaneous arteries. The fascial, the subcutaneous, and the subdermal (cutaneous) layers are shown. Illustrated after Scharnack. (From Daniel and Kerrigan, 1990.)

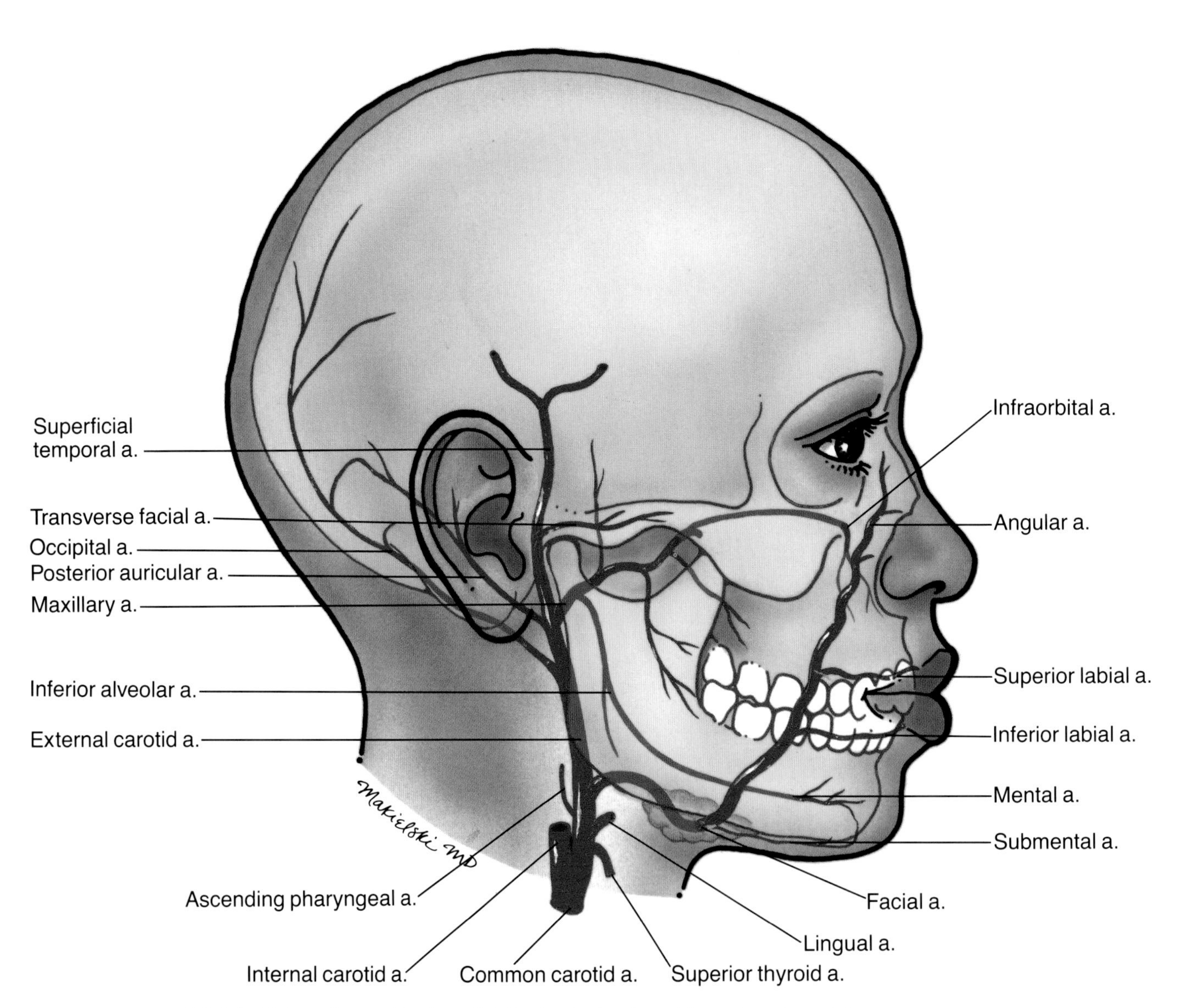

FIG. 9.3. External carotid artery distribution in the face. The external carotid artery separates in the neck from the common carotid artery, and, after giving off the superior thyroid, the lingual, and the ascending palatine arteries, divides into the facial artery and a posterior division that terminates in the superficial temporal artery.

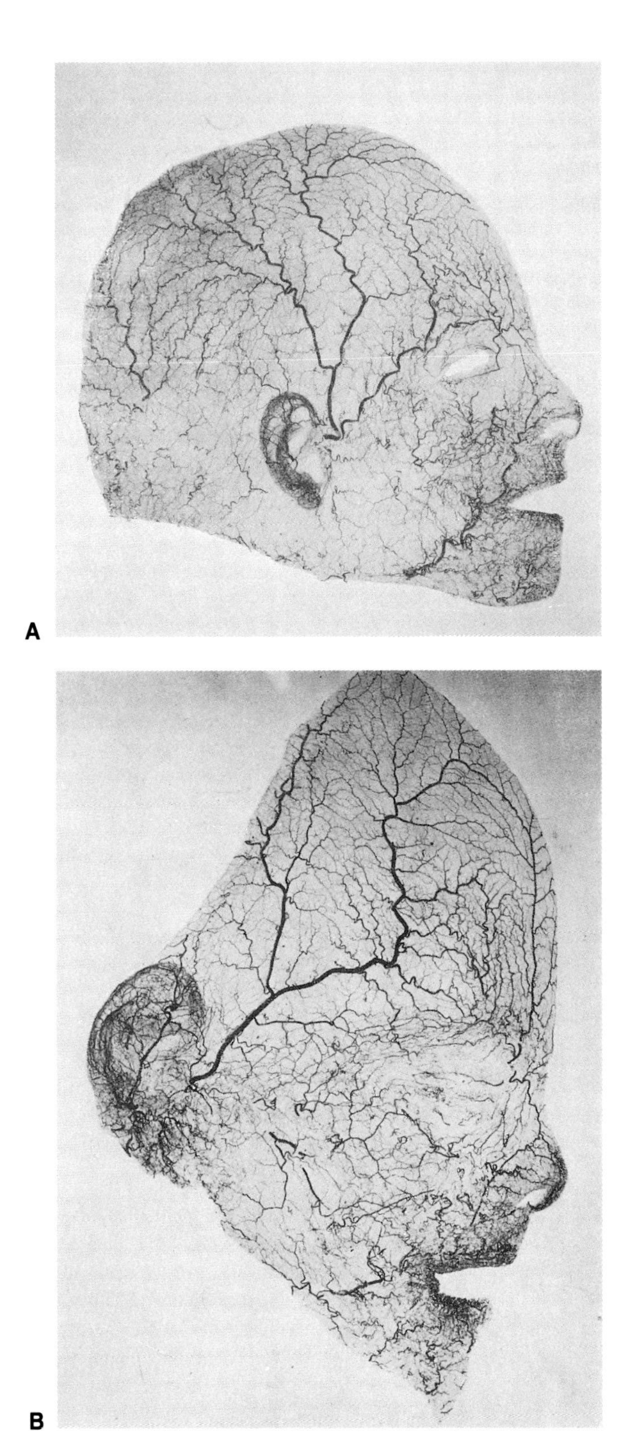

FIG. 9.4. A, B: Vascular patterns of the face. The superficial temporal and the facial vessels anastomose with the supraorbital, the supratrochlear, and the infratrochlear vessels in the mid-face. (From Salmon, 1988, with permission.)

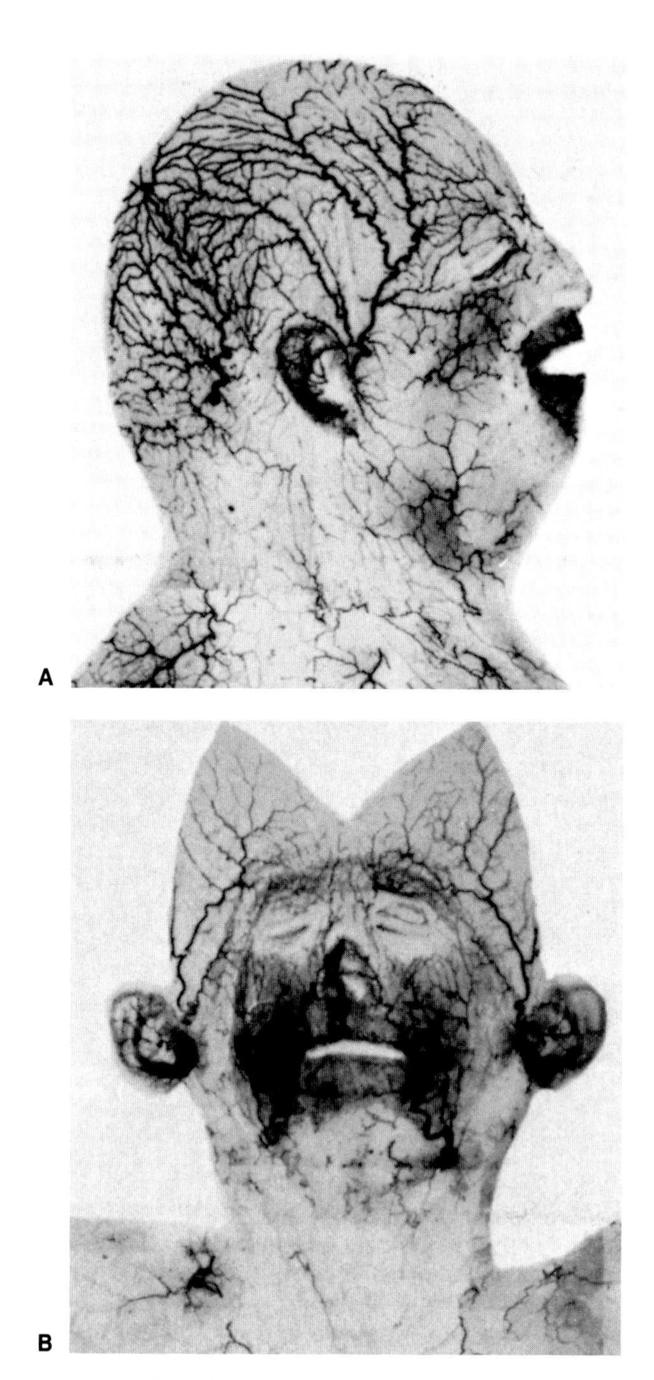

FIG. 9.5. Vascular supply of the face. A, B: Frontal and lateral views of the facial vasculature. (From Taylor and Palmer, 1987, with permission.)

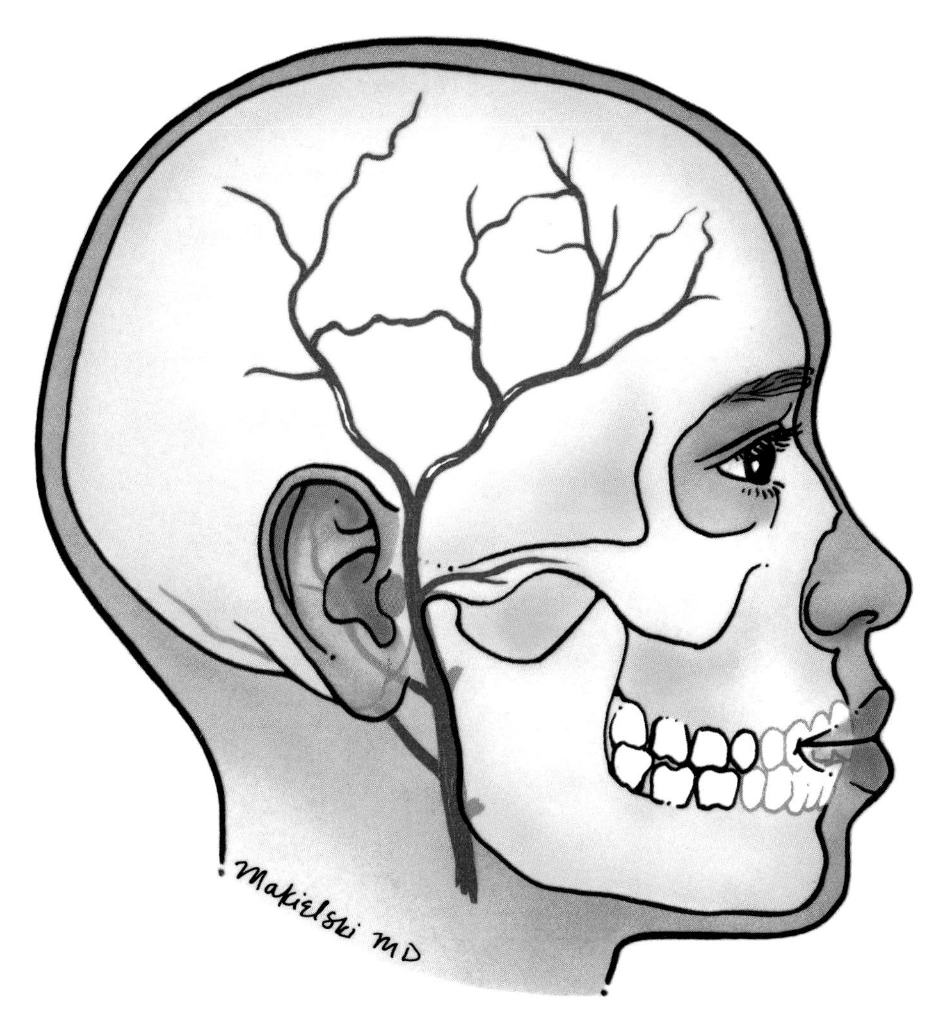

FIG. 9.6. Superficial temporal artery distribution. The most common branching pattern is demonstrated with a bifurcation into the anterior and the posterior branches. Inferior to the zygomatic arch, the transverse facial artery is seen.

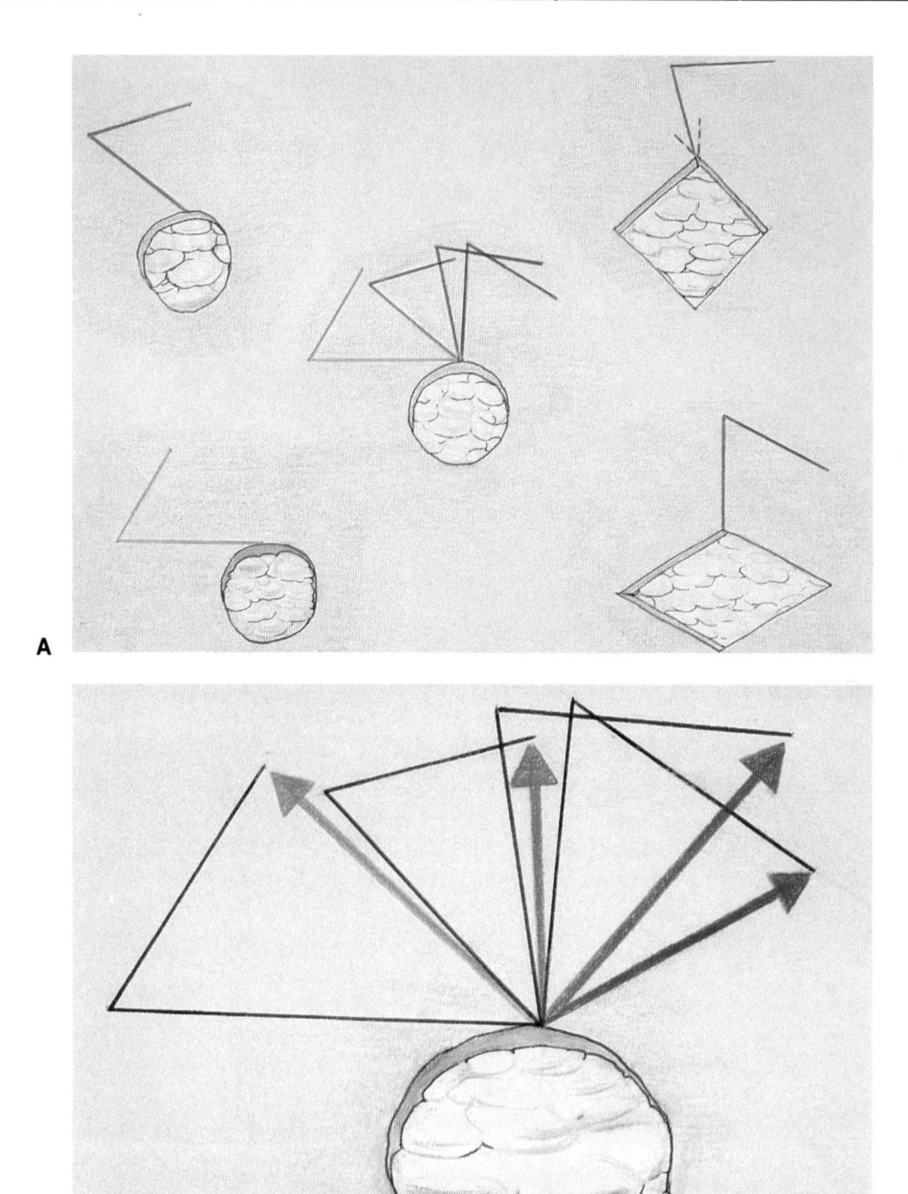

FIG. 9.7. The design of local transposition flaps. Local skin flaps with a blood supply based on the subdermal plexus are the reconstructive technique of choice for repair of most medium-sized facial defects. A large number of specific geometric flap designs have been described, but transposition flaps are the most widely used. These flaps share certain characteristics. **A:** When closing a circular defect with a transposition flap, a number of designs are available. These include (clockwise from the top left): a 60 transposition flap, a Dufourmentel flap, a Limberg rhombic flap, and a "note" flap. In the center of the figure these flaps are superimposed on the circular defect. For each flap the sides are approximately the length of the defect's diameter and the tip angle is approximately 60°. The major variables thus become (1) the site of flap origin along the defect circumference and (2) the angle the initial flap side makes with a tangent to the defect; this varies from 0° for the "note" flap to 90° for the Limberg rhombic flap. **B:** The major tension in each flap is at the closure of the donor sites. This tension vector is demonstrated for each flap by the *arrows.* The anatomic problem in transposition flap design thus becomes creating an individualized transposition flap that places these tension vectors in the lines of maximal extensibility (LME) and the donor site closure in the relaxed skin tension lines (RSTL) as best as possible. Obviously, the need to distribute tension to avoid distortion of anatomic landmarks, such as the eyelid, takes precedence over these theoretical considerations. Illustrated by Joanne Clifford. (From Larrabee, 1992. Copyright 1992, American Medical Association.)

CHAPTER 10

Lymphatics of the Face

All regions of the head and the neck, with the exception of the deep orbit, contain a rich supply of lymphatics. Of the estimated 800 lymph nodes in the body, approximately 300 are located in the head and neck region. Many tumors and infections of the head and the neck spread along the lymphatic drainage system.

Lymphatic capillaries form a rich anastomotic plexus in the facial tissues and along the upper aerodigestive tract. These vessels coalesce into channels that lead to nearby lymph nodes. Ultimately, most of the lymphatics in the head and the neck drain to the deep cervical nodes. From there, the lymph enters the venous system by way of the thoracic duct on the left and the lymphatic duct and the lower internal juglar vein on the right. The collected lymph often passes through a series of nodes en route to joining the venous system. As a result, all lymph from the head and the neck filters through at least one and usually several lymph nodes prior to entering the venous circulation.

LYMPHATIC DRAINAGE OF THE FACE

Lymphatics in the face generally follow the venous drainage, coursing inferiorly and posteriorly. The facial lymphatics can be divided into three major patterns (Fig. 10.1). The midfacial tissues drain into lymphatics that generally follow the facial vessels, terminating in the facial, the submental, and the submandibular lymph nodes. The lateral face and the frontotemporal scalp drain along a posterior and diagonal line into the parotid lymph nodes. Finally, the parietal and the occipital scalp drain away from the vertex of the skull: the parietal scalp drains to the parotid lymph nodes anteriorly and the retroauricular (mastoid) lymph nodes posteriorly, whereas the occipital scalp drains posteriorly into the occipital group of lymph nodes.

These regional lymphatic patterns form a band of lymphatic tissue that borders the junction of the head and the neck and serves as a first echelon filtering system for the face, the scalp, and the mucous membranes (Fig. 10.2). Rouviere aptly named this drainage system the "pericervical" group of lymph nodes. Together with the retropharyngeal lymph nodes, the pericervical system forms a ring of lymphoid tissue that surrounds the upper aerodigestive tract and base of skull.

Midface

The central face drains by way of the facial nodes to the submental and the submandibular nodes and, to a lesser degree, the periparotid nodes. The facial group of nodes drain the medial eyelids, the medial cheek, the external nose, and the upper lip. Approximately five to ten facial nodes are located in the subcutaneous tissues along the course of the facial vessels. Most of the facial nodes drain directly into the submandibular region.

The lower lip, the chin, the anterior tip of the tongue, and the floor of the mouth drain in a straight-line pattern down to the submental lymph nodes. The submental nodes also receive input from the medial cheek and portions of the gingiva. The submental lymph nodes number from one to eight and are found in the fatty tissues deep to the platysma and superficial to the mylohyoid muscle, between the anterior bellies of the digastric muscles. The submental region drains into the submandibular lymph nodes.

The submandibular lymph nodes total from three to six and lie close to the submandibular gland and the facial vessels. Drainage from the submandibular region is into the supraomohyoid portion of the internal jugular lymph node chain.

Nose

The lymphatics from the nasal tip and the dorsum follow the venous drainage and drain laterally into the submandibular nodes. There is essentially no lymphatic drainage through the columella (Fig. 14.13).

Lateral Face and Temple

Lymphatic drainage from the lateral face and the temple occurs in a posterior and inferior direction toward the parotid gland. Two groups of lymph nodes are found in association with the parotid gland: the intraglandular and the extraglandular (paraglandular or preauricular) lymph nodes. The parotid lymph nodes number from 20 to 30 and drain into the internal jugular lymphatic chain.

Because the gestational development of the lymphatic system precedes the formation of the parotid gland, lymph nodes may be trapped in the substance of the developing gland parenchyma. Virtually all intraglandular nodes lie lateral to the posterior facial vein; however, their relationship to the facial nerve is variable. Intraglandular nodes receive input from the lateral eyelids, the temporal and the parietal scalp, the forehead, the lacrimal gland, the conjunctiva, the external auditory canal, the eustachian tube, and the parotid gland itself.

The extraglandular nodes receive lymphatic input from the forehead and the temporal scalp, the external auditory canal, the anterior pinna, the lateral upper and the lateral lower eyelids, and the root of the nose. Paraglandular nodes that lie along the tail of the parotid receive input from the cheek and the buccal mucosa, the pinna, and the parotid gland itself.

Ear

The lymphatic drainage of the auricle is split based on its embryology. The tragus and the helical root, which derive from the first three hillocks, drain anteriorly. The remainder of the auricle, from the fourth, fifth, and sixth hillocks, drains into the postauricular lymph nodes. This drainage pattern is of particular importance in management of malignancies of the auricle.

Parietal and Occipital Scalp

Drainage from the parietal scalp may be divided into an anterior and a posterior pathway. Anterior drainage is into the parotid lymph nodes, whereas posterior drainage is into the retroauricular lymph nodes. The retroauricular nodes, which number from one to four, overlie the mastoid cortex and drain either to the spinal accessory group or the internal jugular chain. Drainage from the occipital scalp passes into the occipital lymph nodes. These lymph nodes drain primarily into the spinal accessory chain.

LYMPHATIC DRAINAGE OF THE NECK

From the pericervical region, lymph flows into the deep system of the lateral cervical lymphatics. The lateral cervical lymph nodes are commonly divided into three major groups, based on their proximity to surrounding neurovascular structures: the internal jugular, the spinal accessory, and the transverse cervical lymph nodes. These three groups form an approximate triangle on the lateral neck (Fig. 10.3). An anterior cervical group of lymph nodes is also found in the neck; however, these nodes are less important in drainage of the facial tissues.

Internal Jugular Nodes

The internal jugular group, which generally follows the course of the jugular vein, consists of 15 to 40 lymph nodes. Lymph flows to the internal jugular system from the submental, the submandibular, the parotid, the retroauricular, and the retropharyngeal lymph nodes. Most of the nodes lie between the posterior belly of the digastric muscle and the omohyoid muscle. A large jugulodigastric or "tonsillar" lymph node marks the level where the posterior belly of the digastric muscle crosses the jugular vein. The jugulodigastric node is frequently palpable when involved with disease. It receives lymphatic input from the submandibular nodes, the tonsil, and the oropharynx.

Another important lymph node in the internal jugular system is the juguloomohyoid, which marks the transition between the upper and the lower internal jugular chain. The most inferior nodes in the internal jugular chain, the nodes of Virchow, often receive metastatic implants from tumors originating in the abdominal and the thoracic cavities.

Spinal Accessory Nodes

The spinal accessory nodes follow the course of the spinal accessory nerve. There are four to 20 lymph nodes in this group, which drain the occipital and the retroauricular lymph nodes, the parietal and the occipital scalp, and the skin overlying the lateral and the posterior neck. At the superior aspect of the spinal accessory chain, "junctional" nodes lie in close proximity to the uppermost nodes of the internal jugular system.

Transverse Cervical Nodes

The transverse cervical (supraclavicular) lymph nodes form the third leg of the deep lateral system, connecting the accessory chain laterally with the jugular chain medially. Primary afferent input is from the spinal accessory chain; however, the transverse cervical system also receives lymphatic drainage from the breast and the anterior thoracic wall. The transverse cervical lymphatics drain into the inferior aspect of the internal jugular system.

Anterior Cervical Nodes

An anterior cervical group of lymph nodes is also found in the neck; however, these nodes are less important in drainage of the facial tissues than the three groups discussed above. The anterior cervical nodes are classified into two groups: the superficial (anterior jugular) and the juxtavisceral nodes. The anterior jugular nodes provide variable drainage for the skin and the muscle that overlie the anterior neck region. The juxtavisceral nodes primarily drain the visceral compartment of the neck. Lymph from this region proceeds to the inferior jugular nodes.

CLASSIFICATION SYSTEMS

Many classification systems have been introduced to organize the head and the neck lymphatics, most borrowing from the early work of Rouviere. A recent classification system (Shah et al., 1981) divides lymph nodes into seven major levels based on a clinically oriented anatomic scheme (Fig. 10.4).

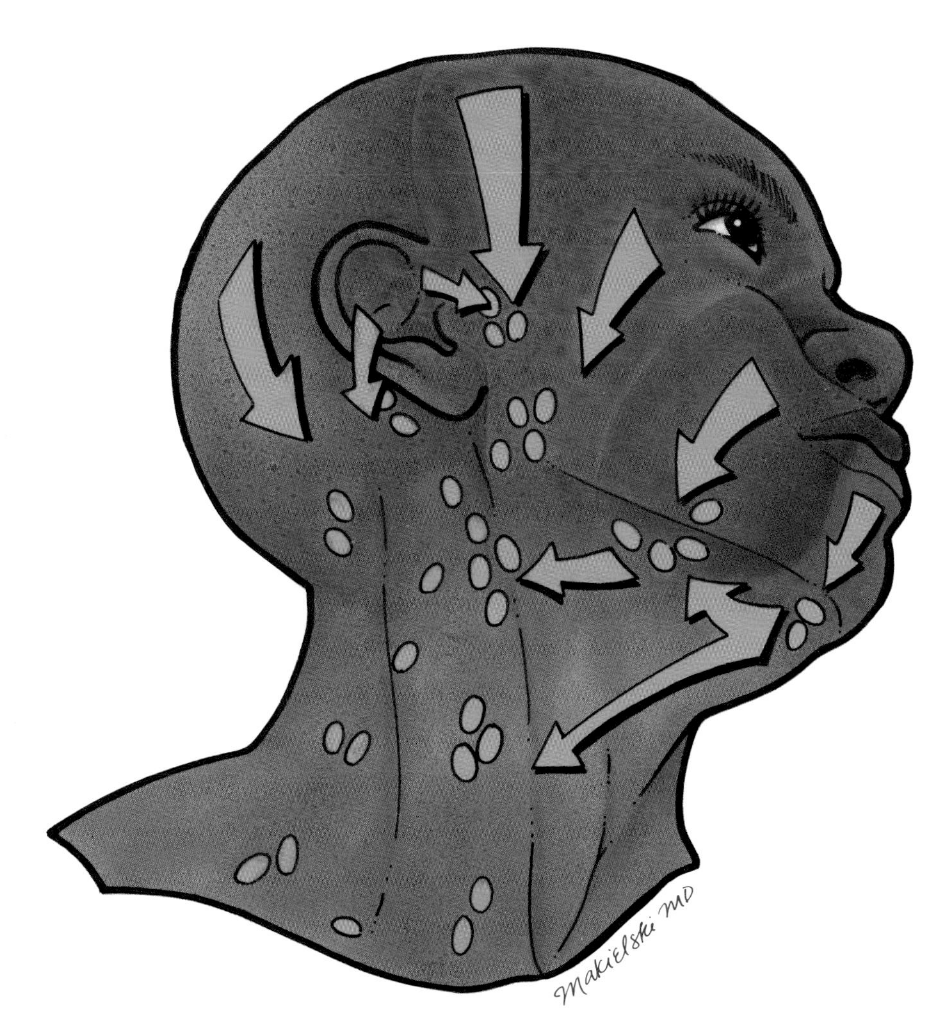

FIG. 10.1. General pattern of lymphatic drainage of the face and the scalp. The midface drains into lymphatics that follow the facial vessels to the facial, the submental, and the submandibular nodes. The lateral face and the frontal scalp drain to the parotid nodes. A line drawn from the lateral canthus to the angle of the mandible defines approximately the division between the anterior and the lateral facial drainage patterns. The posterior drainage of the scalp forms a third regional pattern.

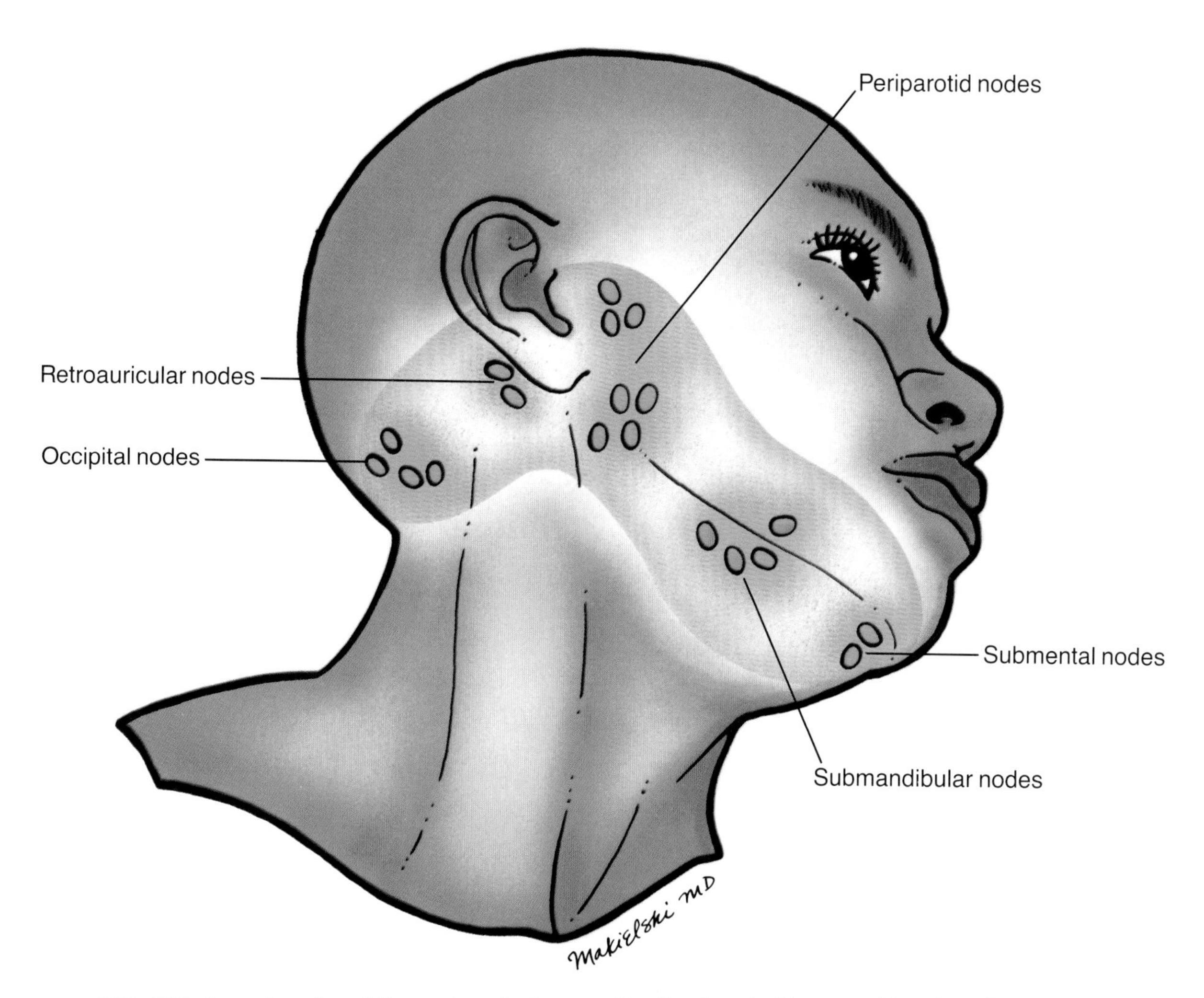

FIG. 10.2. Lymph nodes of the pericervical group. The "pericervical" group of lymph nodes borders the junction of the head and the neck (Rouviere). This "ring" of lymph nodes serves as a first-echelon filtering system for the face, the scalp, and the mucous membranes.

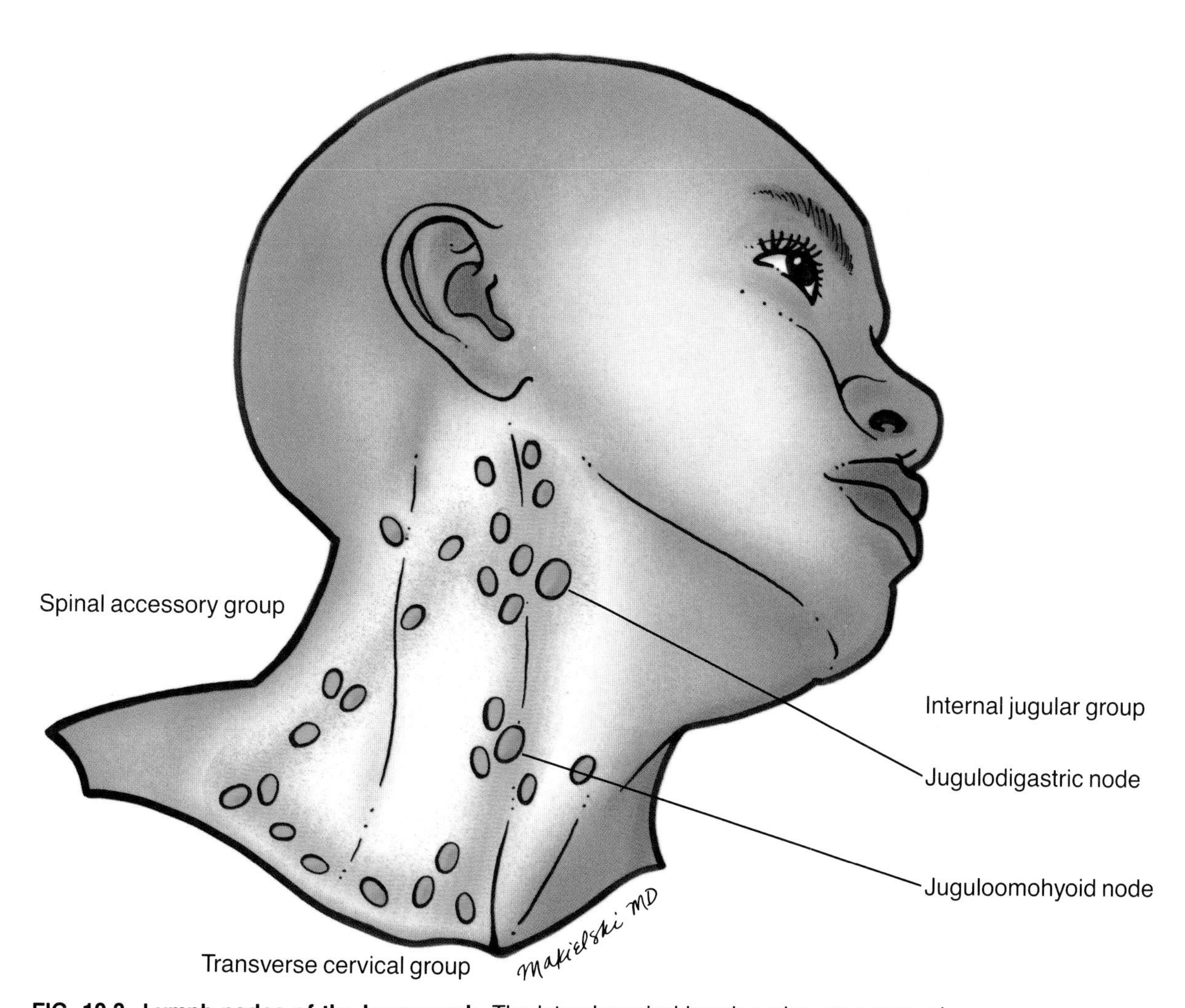

FIG. 10.3. Lymph nodes of the lower neck. The lateral cervical lymph nodes are commonly divided into three major groups: the internal jugular, the spinal accessory, and the transverse cervical lymph nodes. At the superior aspect of the neck; "junctional" lymph nodes of the spinal accessory chain lie in close proximity to the uppermost internal jugular nodes. The transverse cervical nodes connect the accessory chain laterally with the jugular chain medially.

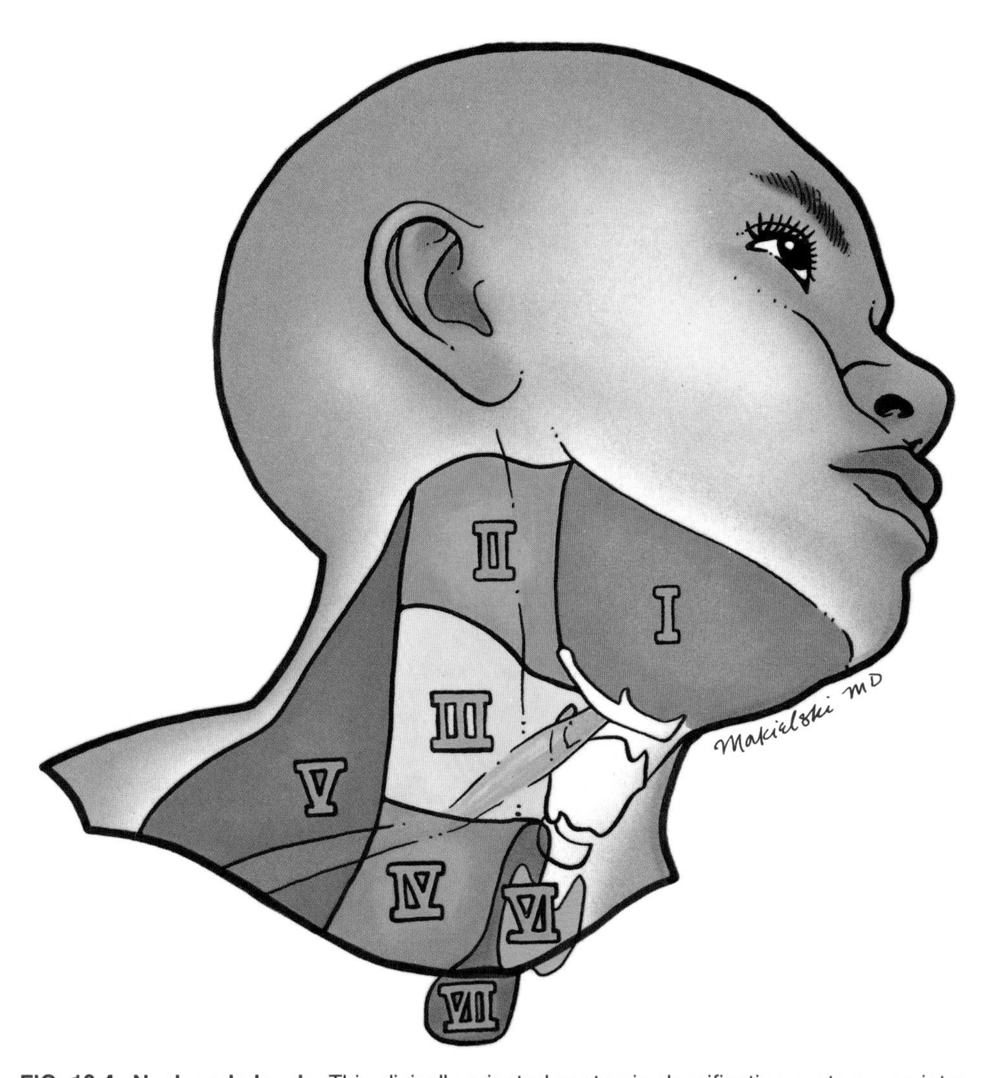

FIG. 10.4. Neck node levels. This clinically oriented anatomic classification system was introduced in 1981 by Shah et al. The submental and the submandibular lymph nodes comprise level I. The internal jugular chain lymphatics are divided into three levels by the hyoid bone (or carotid bifurcation) and the cricoid cartilage (or omohyoid muscle): the superior jugular nodes are classified as level II; the middle nodes, level III; and the inferior nodes, level IV. The accessory lymph nodes are classified as level V. Nodes intimately associated with the thyroid gland are classified as level VI, whereas nodes along the tracheoesophageal groove and in the superior mediastinum comprise level VII.

SECTION III

Anatomic Regions

CHAPTER 11

Hair and Scalp

There are two basic types of human hair: terminal hair, such as is found on the scalp, and vellus hair, which is the short, soft type found over the body. This section is concerned with the terminal hair and the scalp.

HAIR AND SCALP ANATOMY

The hair follicle consists of the hair itself and the surrounding epithelial covering, which includes the sebaceous gland and the arrector pili muscle (Fig. 11.1). The base of the follicle extends to the subcutaneous fat and can be injured easily when elevating skin flaps.

The scalp has a unique anatomy. The skin itself is quite thick. Immediately beneath the epidermis and the dermis lies the firm, dense, and vascular subcutaneous tissue (Fig. 11.2). Because of the multiple fibrous septae, the vessels do not contract well and the scalp bleeds profusely when cut. Beneath the subcutaneous tissue lies the galea aponeurotica, a thick, tough, fibrous tissue that connects the frontalis muscle and the occipital muscle (Fig. 11.3). The complex of these two muscles and the galea is termed the epicranium; it inserts laterally into the temporalis (Fig. 11.4). Beneath the galea is a plane of loose areolar tissue that is relatively avascular. Next lies the pericranium or the periosteum of the skull. The deepest layer is the outer cortex of the skull.

HAIR GROWTH PATTERNS

Hair Growth Cycle

Hair grows in a cyclic manner with periods of growth (anagen) alternating with periods of rest (telogen). The normal anagen phase of the scalp is several years and the telogen phase is 3 to 4 months. The growth cycle can be interrupted by trauma such as hair transplantation or excessive tension in wound closure. The bulb at the lowest end of the follicle contains the undifferentiated cells that develop into the various layers. Damage above this region is usually not permanent and regrowth will occur.

Direction of Hair Growth

The direction of growth of the hair follicles varies over the scalp and is of surgical importance. Since the follicles are not perpendicular to the skin surface but meet it at an angle, the surgeon must make an incision parallel to the follicles to avoid injury. The same principle applies to incisions around the eyebrow or when harvesting punch grafts. There is individual variation in the follicle direction; a usual pattern is shown in Figure 11.5. The follicles in the crown are oriented in a swirling pattern.

Male Pattern Baldness and Hair Growth Patterns

A standard classification of the progression of hair loss in male pattern baldness is presented in Figure 11.6. The male hairline has a distinct temporal recession and temple point (Figs. 11.7 and 11.8). As male pattern hair loss develops, this recession becomes more marked, the temple point is lost, and hair loss develops on the crown.

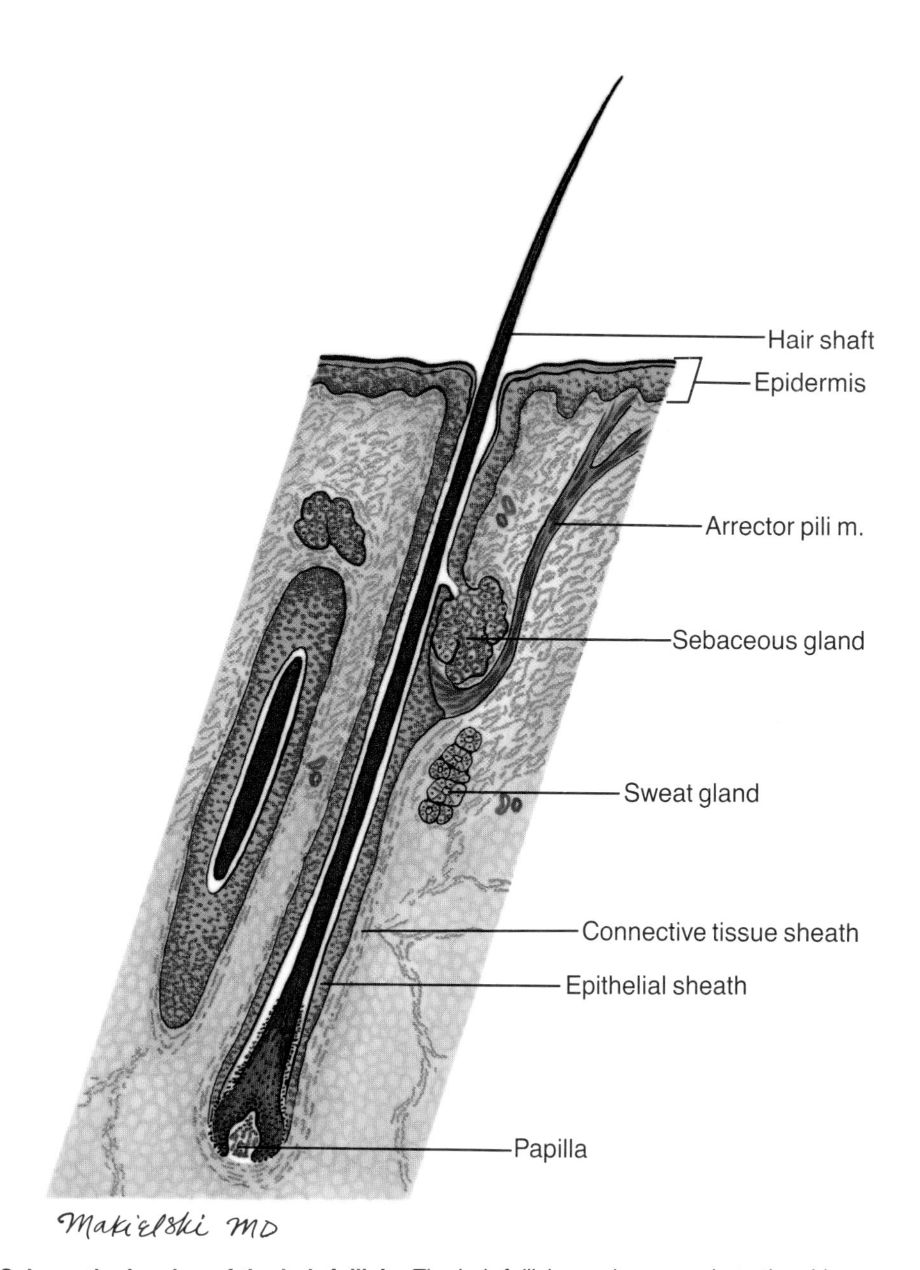

FIG. 11.1. Schematic drawing of the hair follicle. The hair follicles make an angle to the skin and thus can be damaged when skin incisions are made perpendicular to the skin rather than parallel to the hair follicles. The hair follicles lie in the subcutaneous tissue deep to the dermis and can also be damaged when elevating flaps. Conversely, one can preserve them and advance them beneath a beveled incision to obtain future hair growth through a scalp incision. The cells of the hair follicle grow and differentiate from the bulb toward the surface. The arrector pili muscle can be seen to attach beneath the sebaceous gland.

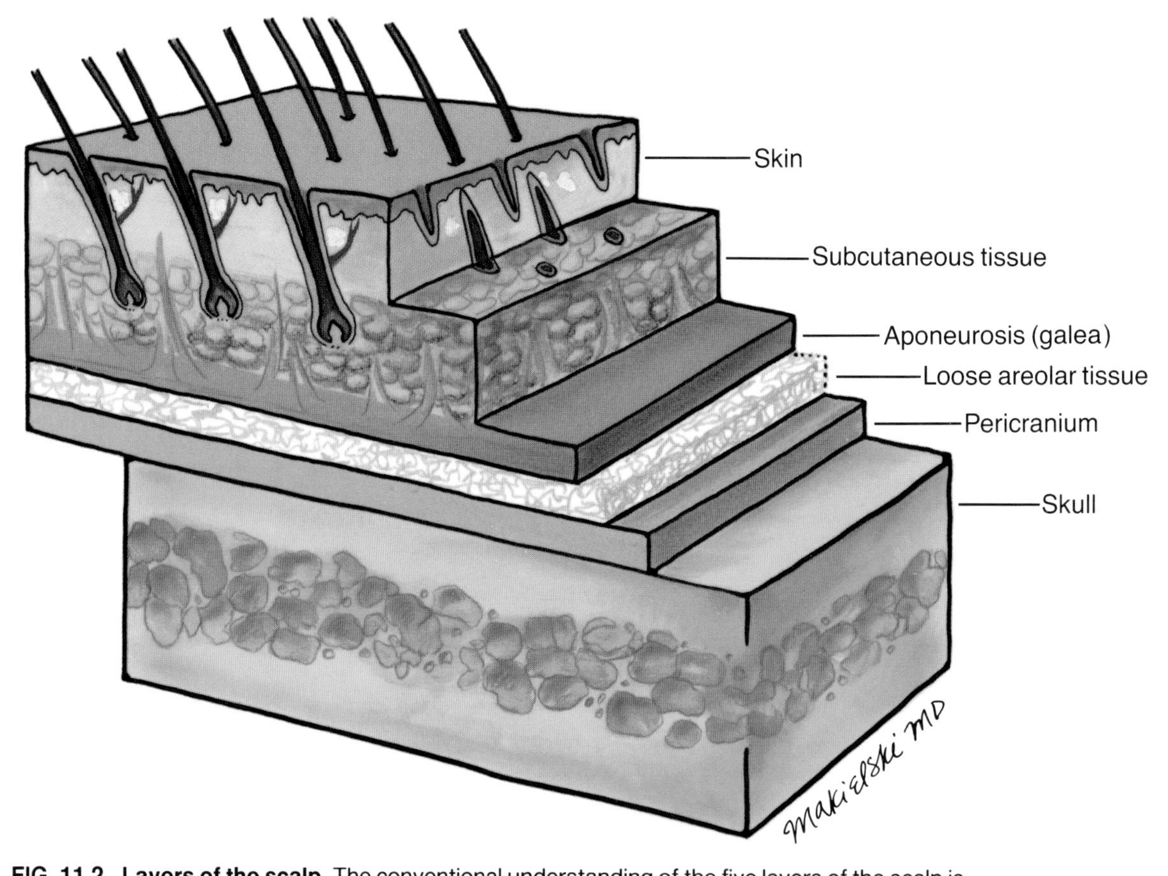

FIG. 11.2. Layers of the scalp. The conventional understanding of the five layers of the scalp is reflected in the time-honored mnemonic:

S = Skin

C = Subcutaneous tissue

A = Aponeurosis and muscle (galea)

L = Loose areolar tissue

P = Pericranium

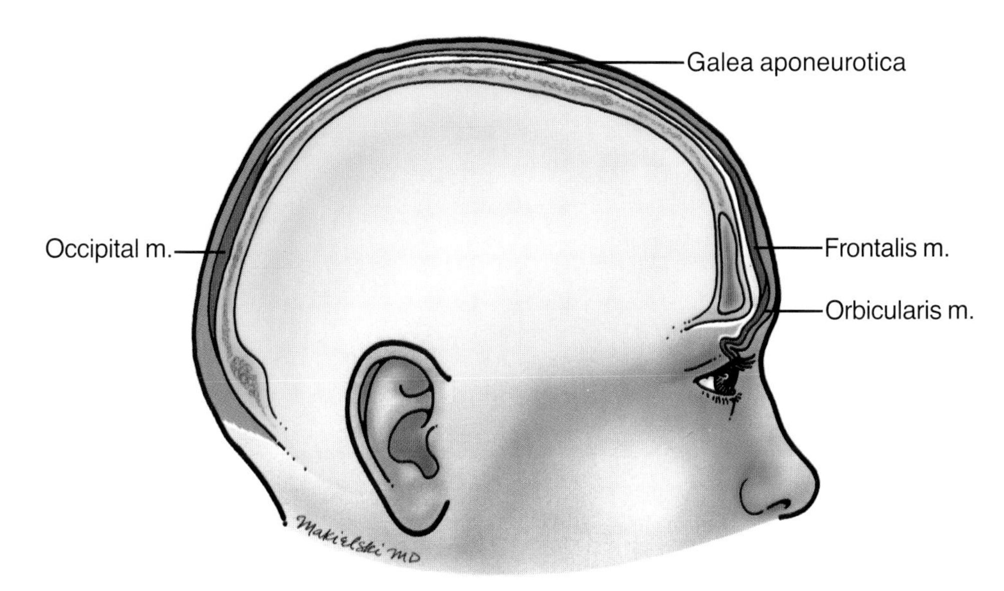

FIG. 11.3. Section through the epicranium. A sagittal section through the epicranium demonstrates the tendinous galea aponeurotica between the frontalis and the occipital muscles.

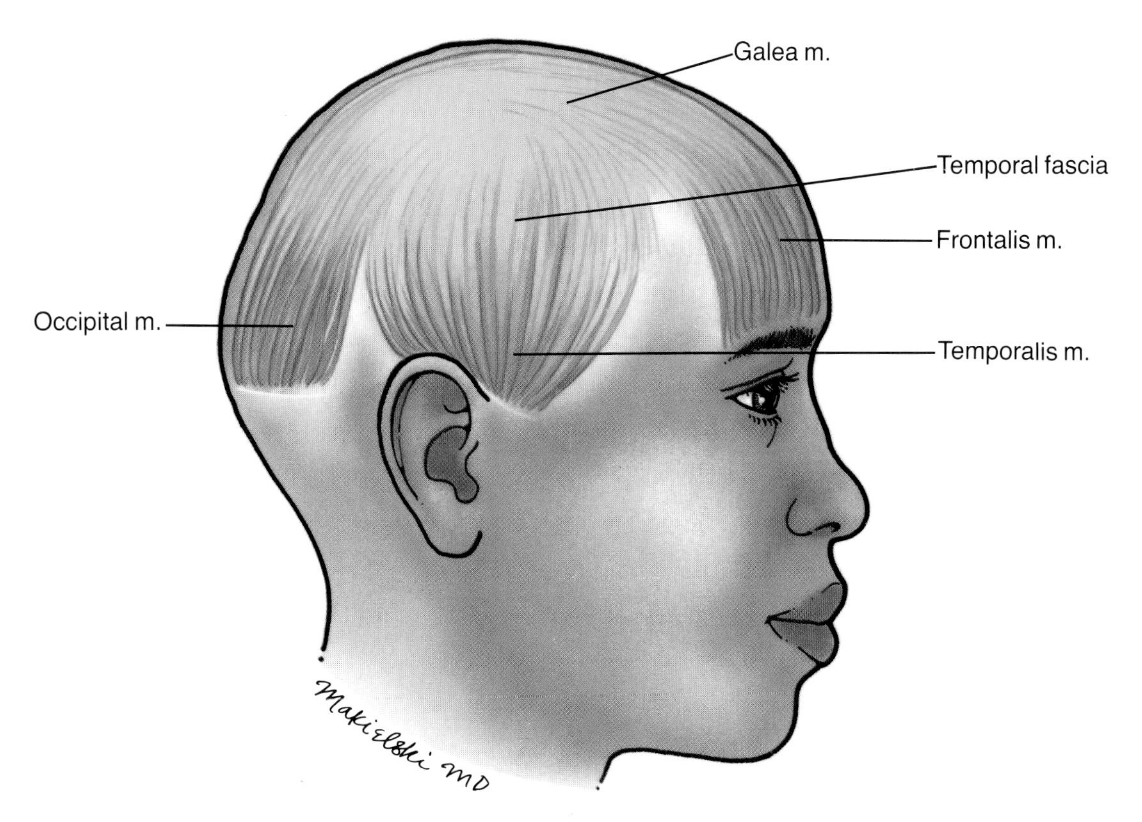

FIG. 11.4. Diagram of the epicranium. The epicranium is composed of the frontalis, the galea, and the occipital muscles; laterally it extends into the temporal parietal fascia (also SMAS or superficial temporal fascia). The frontalis muscle joins the orbicularis muscle at the orbital rim.

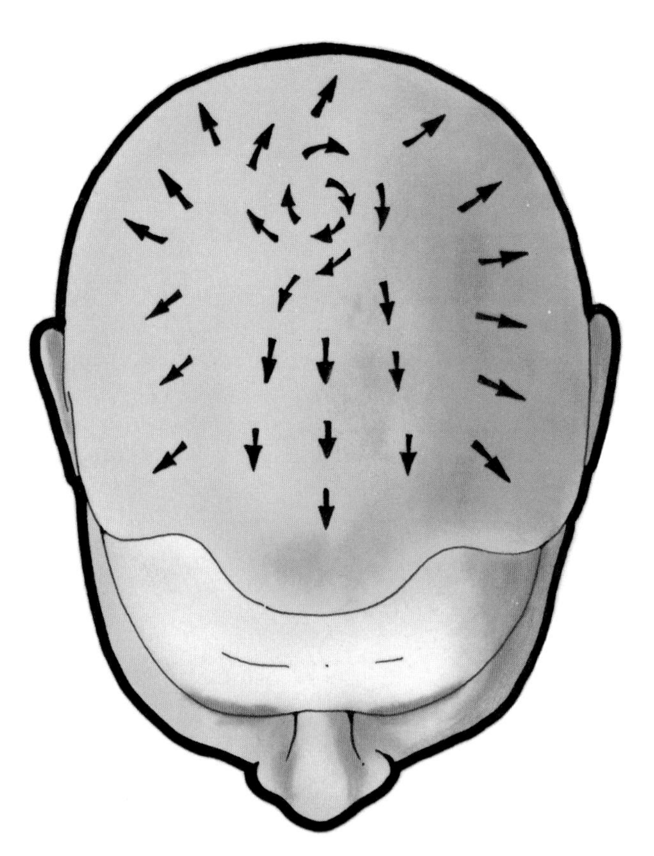

FIG. 11.5. Direction of the hair follicles in the scalp. Incisions beveled with the hair follicles will preserve them. In areas where the direction of hair shafts changes, as in a coronal incision, special care is needed to align the incision with the hair shafts throughout the incision.

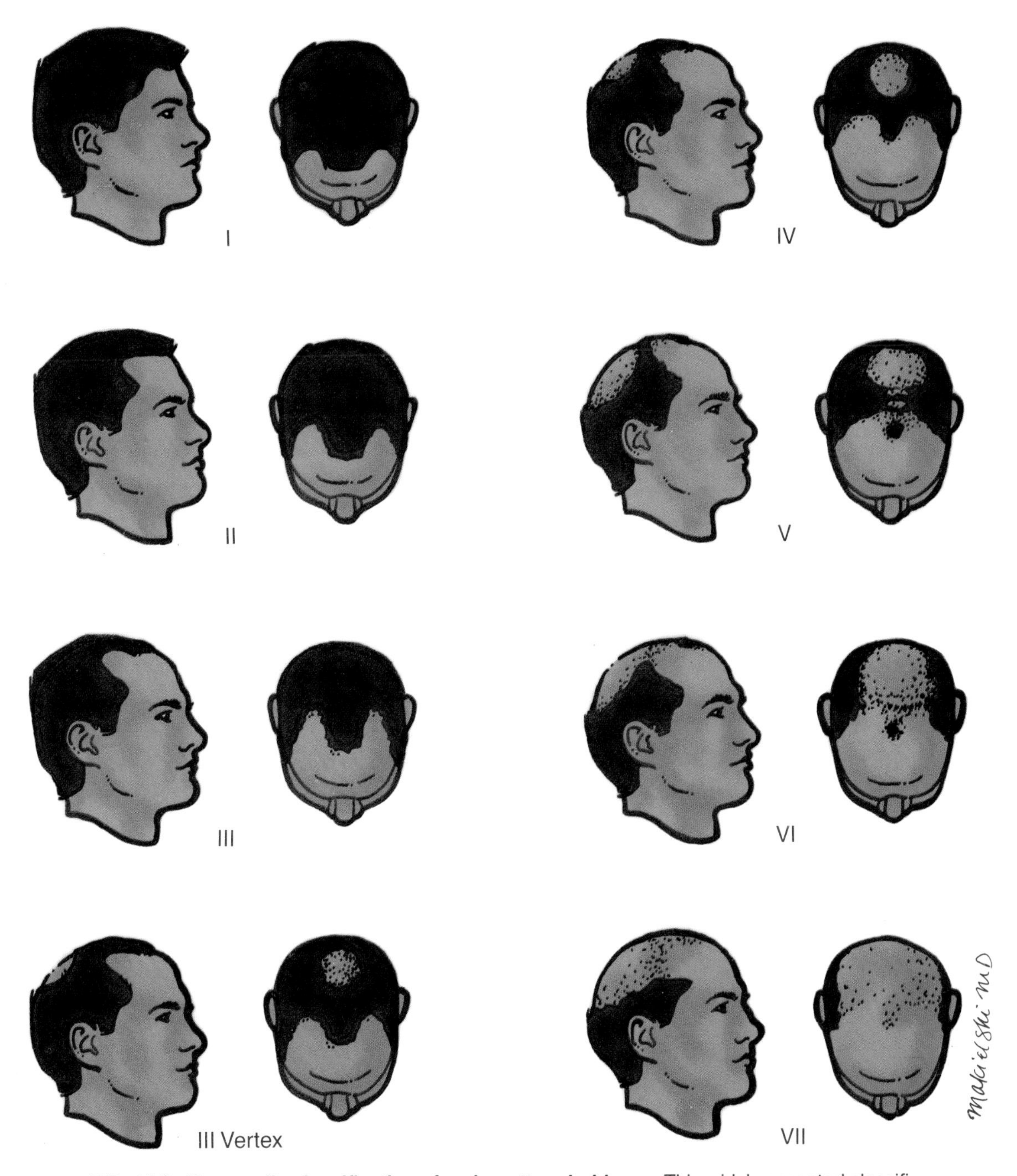

FIG. 11.6. Norwood's classification of male pattern baldness. This widely accepted classification of male pattern baldness standardizes preoperative planning and analysis of results and also facilitates communication between physicians. Although the classification may help to select candidates for various procedures such as scalp reduction, hair plugs, or flaps, it is only one of many relevant variables. Age, family history, scalp laxity, hair texture and density, and hair color can all be equally important. (After Norwood, 1973 and 1975, with permission.)

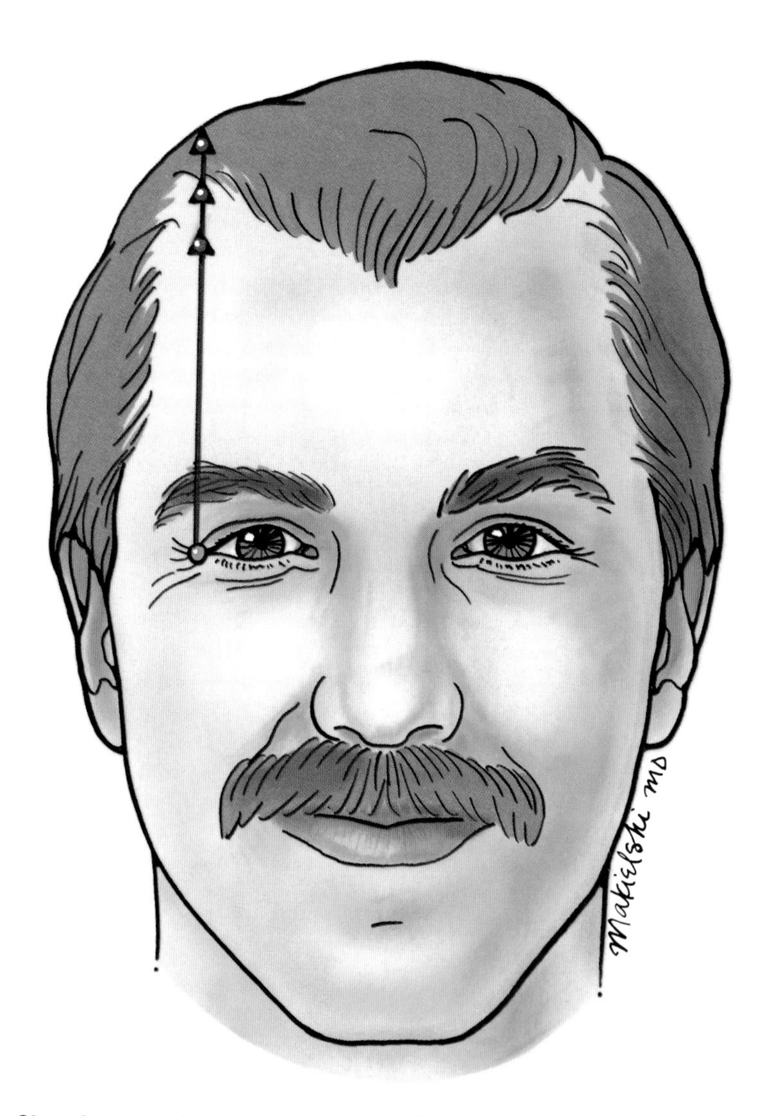

FIG. 11.7. Site of temporal recession in male pattern baldness. The temporal recession of the hairline is in a direct superior line from the lateral canthus, as shown by the arrows in the figure. This relationship is important in designing the hairline for hair replacement procedures.

FIG. 11.8. Anterior position of the hairline. The "rule of threes" is used to determine the appropriate level for the anterior hairline. The distance from the hairline to just below the brow should be the same as the distance from the brow to the base of the nose and from the base of the nose to the chin.

CHAPTER 12

Forehead and Brow

The brow and the forehead comprise an aesthetic unit whose contours are primarily dependent on the shape of: (1) the frontal bone, the supraorbital rims, and the zygoma; (2) the action of the frontalis, the corrugator, and the temporalis muscles; and (3) the characteristics of the skin and the soft tissue.

BROW POSITION

The female brow is usually placed well above the superior orbital rim and is arched with the highest point above the lateral limbus; the male brow is more horizontal and lower (Fig. 12.1). As seen in Figure 12.2, the underlying bone structure, rather than the skin or the soft tissue, can be responsible for aesthetic problems. Brow position can be changed by removing skin and soft tissue at various levels above the brow: at the brow itself, the midforehead, the pretrichal line, or the coronal region.

SURGICALLY RELEVANT STRUCTURES

The coronal approach to the forehead provides unparalleled exposure for reconstructive surgery, trauma surgery, and orbital surgery as well as cosmetic procedures. The flap is lifted in a subgaleal plane above the periosteum medially and immediately above the temporalis fascia laterally. Structures of importance include the supraorbital and the supratrochlear neurovascular pedicles and the frontal branch of the facial nerve. Their positions from this approach are seen in Figures 12.3, 12.4, and 12.5.

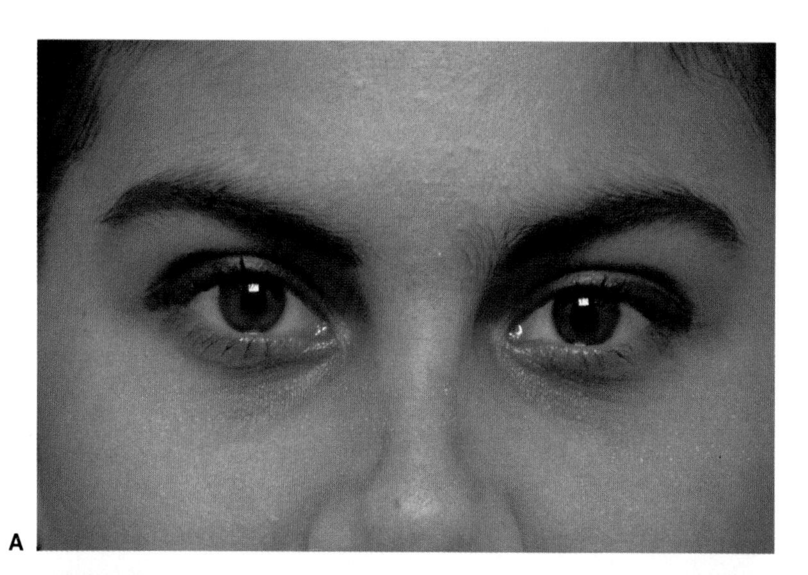

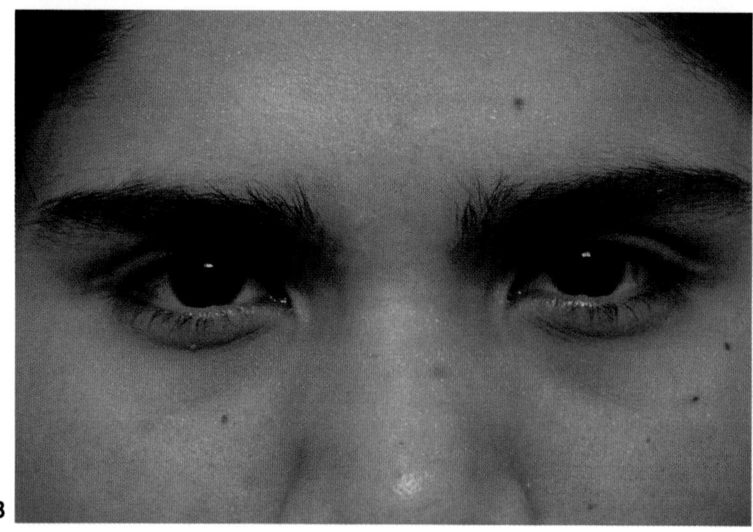

FIG. 12.1. Typical male and female brows. A: The female brow is arched with highest point at approximately the lateral limbus. **B:** The male brow is more horizontal and usually somewhat lower.

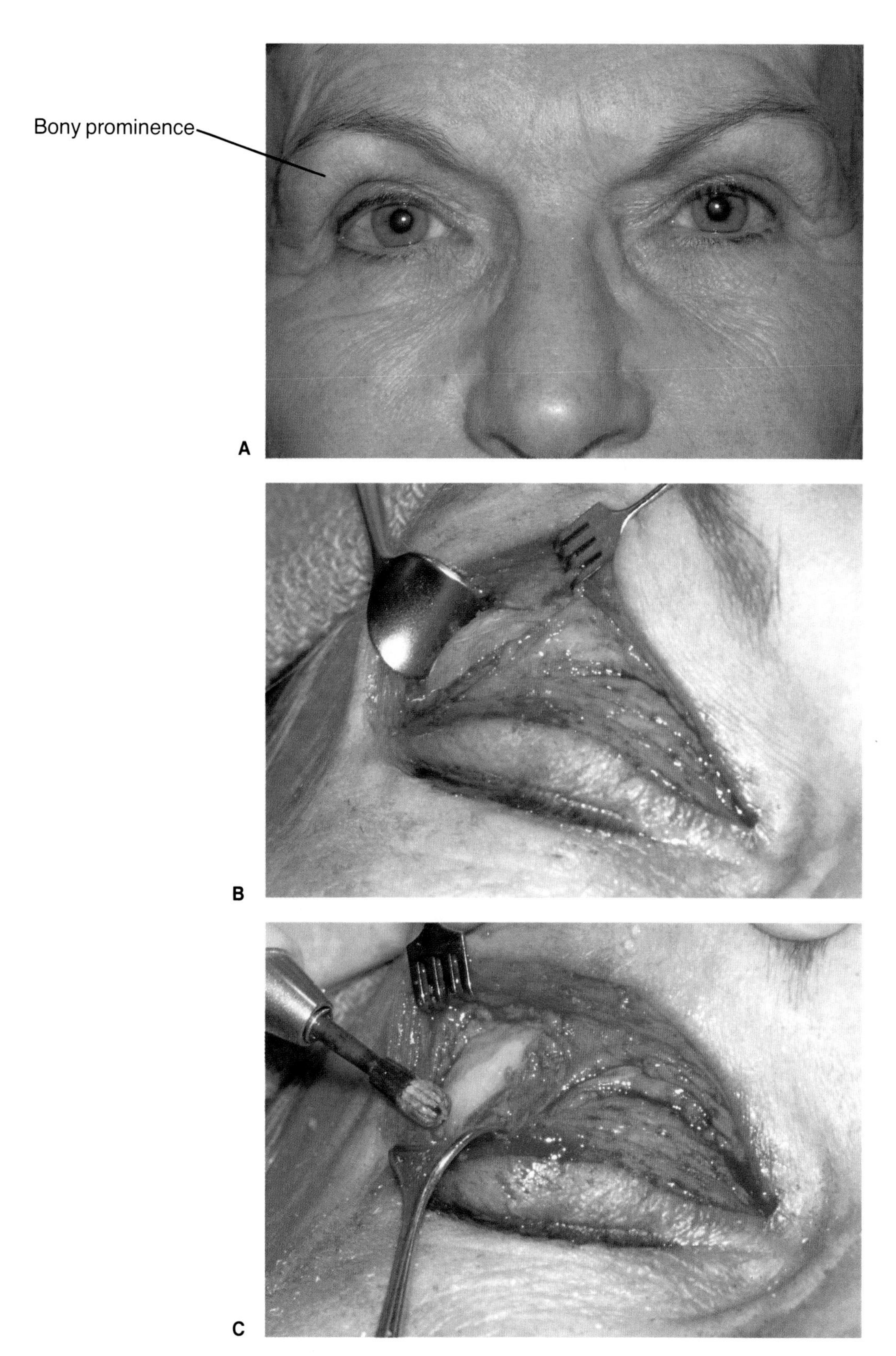

FIG. 12.2. Contour of the orbital bone. The contour of the orbital bone can affect brow aesthetics as much as the soft tissue. **A:** Woman with prominence of supraorbital rims. **B:** Bony rim exposed at surgery through a blepharoplasty incision. **C:** Contouring of the orbital rim.

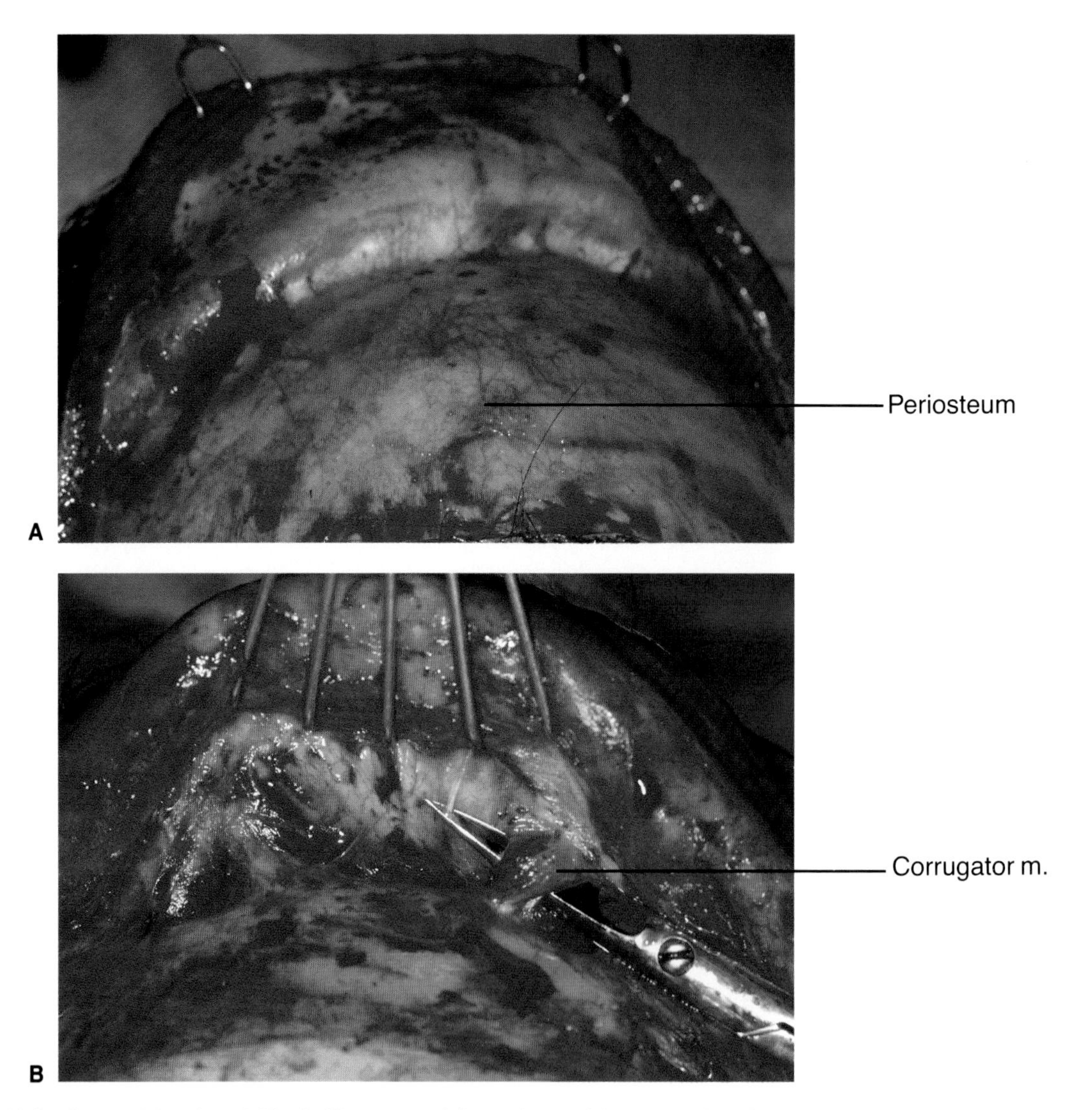

FIG. 12.3. Coronal forehead lift. A: The coronal flap, elevated in a vascular plane above the periosteum, provides excellent exposure to address aesthetic problems of the forehead. **B:** The corrugator muscles may be interrupted to reduce glabellar frown lines.

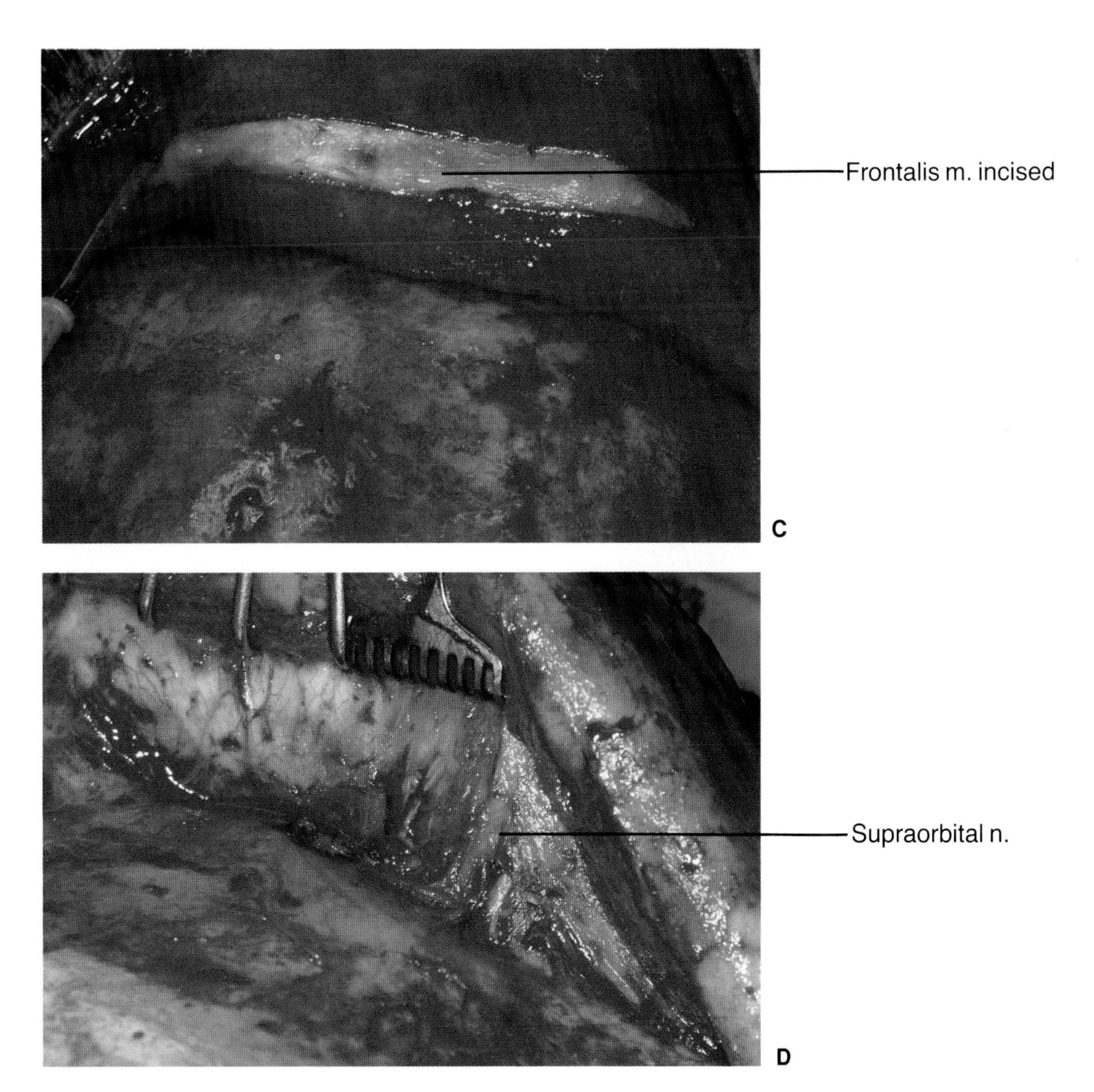

FIG. 12.3. *Continued.* **C:** The frontalis muscle may be partially interrupted in the midline to reduce forehead rhytids. **D:** Important structures, such as the supraorbital nerve, may be easily identified with blunt dissection.

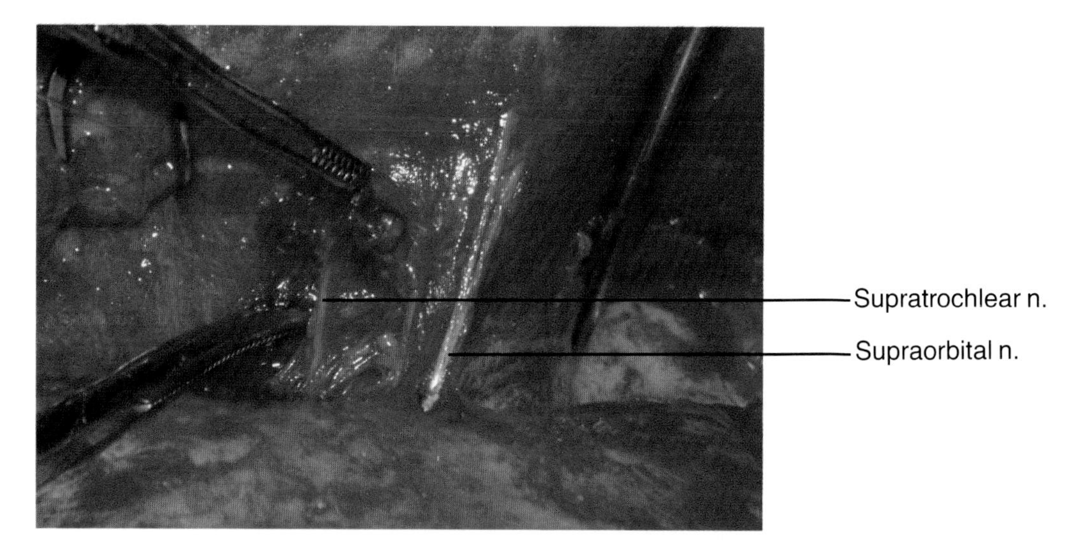

FIG. 12.4. Reconstructive surgery of the orbit via the coronal approach. In this approach the supraorbital and the supratrochlear neurovascular pedicles can be isolated and mobilized by blunt dissection as needed.

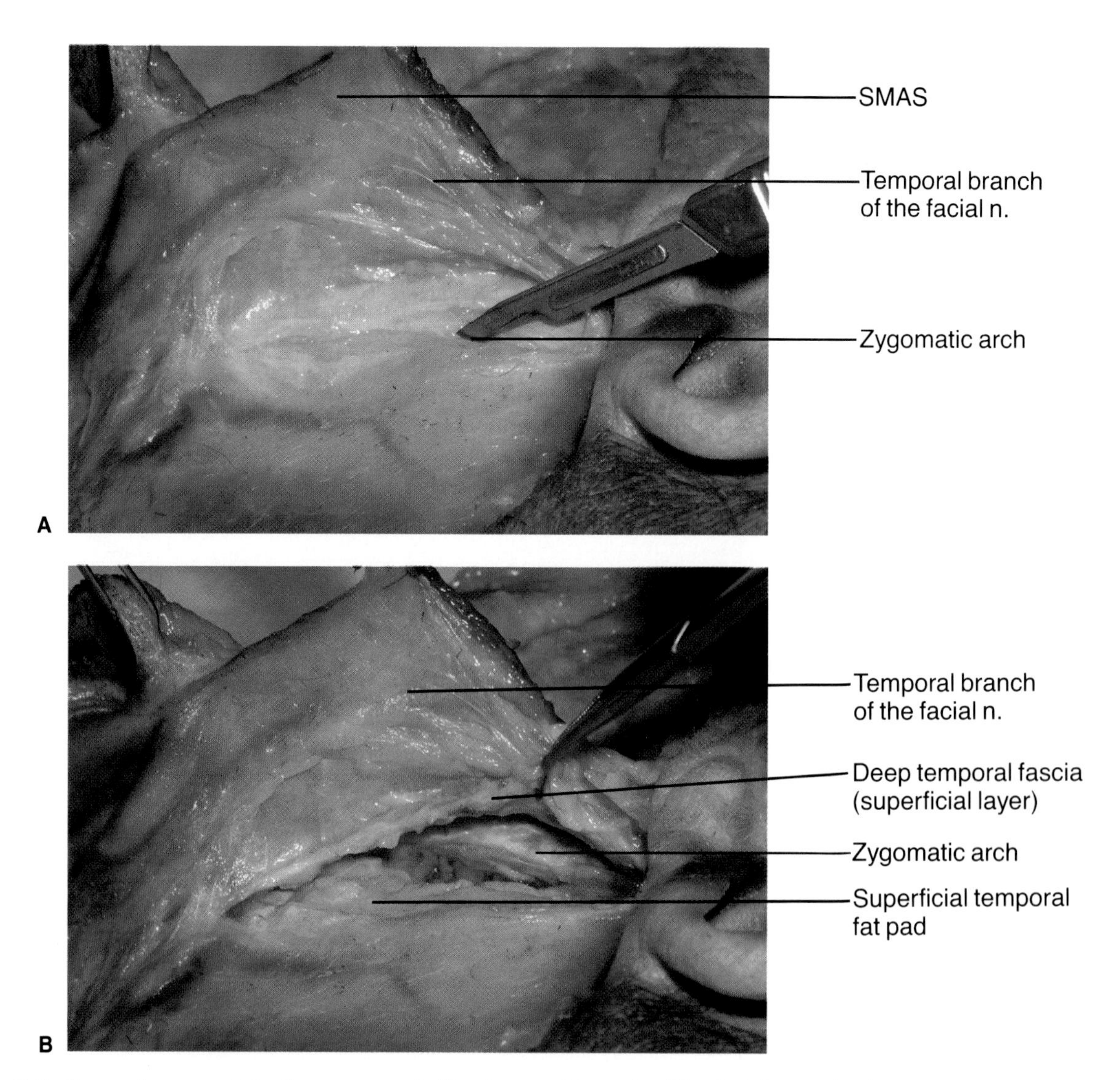

FIG. 12.5. Coronal approach to the zygoma. A: The temporal branch of the facial nerve is at risk as the coronal flap is elevated over the zygoma. **B:** The nerve can be protected when approaching the zygoma by incising the superficial layer of the deep temporal fascia inferior to the plane of the fusion.

CHAPTER 13

Eyelids, Anterior Orbit, and Lacrimal System

An understanding of the anatomy of the orbit is crucial for both cosmetic and reconstructive surgery in this region. This section will emphasize the practical anatomy of the eyelids and associated structures relevant to facial plastic surgery procedures such as blepharoplasty, correction of abnormalities of lid position, lid reconstruction after cancer excision, lacrimal system surgery, and traumatic repairs.

SUPERFICIAL TOPOGRAPHY AND LID ANATOMY

The superficial topography is depicted in Figure 13.1. The eyelid skin is extremely thin and, in youth, quite elastic. The eyelid clearly demonstrates the underlying soft-tissue attachments such as those of the levator aponeurosis. The skin is adherent over the tarsus, but relatively loose over the rest of the lid. There are noticeable differences between Asian eyelids and Caucasian or African eyelids, primarily in the upper lid. The Caucasian and African lids are similar; in the following discussion the term Caucasian will be used to denote all non-Asian eyelids. Within each racial group there are also large differences among individuals; in fact, the anatomy described as typical for the Asian population can occasionally be identified in the Caucasian patient. Typical racial variations in the upper eyelid fold are seen in Figure 13.2.

The contour of the eye is largely determined by the shape of the underlying orbital bone. The relative recession or prominence of the globe with respect to the orbital margins is important both aesthetically and functionally (Fig. 13.3). Proptosis, such as seen in Graves' ophthalmopathy, is measured from the lateral orbital rim to the cornea.

The upper and lower eyelids have analogous structures with similar functions, but the structures are less well defined in the lower lid (Fig. 13.4). The discussion below focuses on the upper eyelid.

Orbicularis Muscle

The orbicularis muscle is arbitrarily divided into an orbital portion, which overlies the orbital margin, a palpebral portion related to the lids, and a lacrimal portion related to the lacrimal pump mechanism (Fig. 13.5). The orbital portion arises from the anterior aspect of the medial canthal tendon and the periosteum above and below it. The palpebral portion of the orbicularis muscle is commonly divided into preseptal and pretarsal parts: the preseptal part overlies the orbital septum and the pretarsal part overlies the tarsal plate. The preseptal portion arises from two heads: a deep head attached to the lacrimal fossa and the posterior crest and a superficial head arising from the medial canthal tendon. Laterally, the preseptal muscles join at the lateral palpebral raphe. The pretarsal part also originates from two heads: a deep head, termed the tensor tarsi (Horner's muscle), and a superficial head from the medial canthal tendon. The superficial heads of the preseptal and the pretarsal muscles in essence become the medial canthal tendon (Fig. 13.6).

Medial Canthal Tendon

The medial canthal tendon is a complex structure that has an intimate relationship to the lacrimal sac. The sac lies between the anterior and the posterior limbs of the medial canthal tendon. Contraction of the tendon creates a pumping action in the sac (Fig. 13.6b). The fascia thickens above the tendon, covering the superior aspect of the sac and creating a "vertical component" of the tendon. The anterior, posterior, and vertical components of the medial canthal tendon complex all provide some support to the medial canthus. The posterior limb is the primary determinant of the medial canthal angle; interruption of only the broadly based anterior limb does not change the position of the medial canthus significantly and can be repaired with a simple suturing of the anterior tendon. Injury of the posterior component, however, causes an anterior displacement of the canthus and must be repaired by wiring the limb to the posterior lacrimal crest.

Lateral Canthal Tendon

Laterally, the pretarsal muscles join to form the lateral canthal tendon, which inserts on the periosteum of the orbital tubercle about 5 mm behind the rim (Fig. 13.7). There is disagreement about the anatomy in this region. Some authors describe a superficial head of the tendon attaching anteriorly to the periosteum of the rim. There is also no consensus as to whether the orbital septum is superficial or deep to the lateral canthal tendon. Although the anatomy of the lateral canthal tendon itself is fairly straightforward, there is disagreement whether it arises from the pretarsal muscle, the tarsal plate, the fascia, or some combination of these. Other structures that attach to the lateral retinaculum along with the lateral canthal tendon include the lateral horn of the levator aponeurosis, the suspensory ligament of the globe (Lockwood's ligament), and the lateral ligament of the lateral rectus.

Orbital Septum and Fat

The orbital septum is a key structure in cosmetic blepharoplasty and an important landmark in functional eyelid surgery such as ptosis surgery (Fig. 13.8). The origin of the septum closely follows the orbital rim except in the medial and the lateral canthal areas. This fibrous structure originates from the periosteum of the orbital margins including the tough arcus marginalis of the frontal periosteum. It lies deep to the orbicularis muscle; in the upper lid it fuses with the levator aponeurosis about 2 or 3 mm above the tarsus (in the non-Asian lid). In the lower lid the septum fuses with the capsulopalpebral fascia, which is analogous to the levator aponeurosis of the upper lid, approximately 5 mm beneath the tarsus. Medially it attaches to the posterior lacrimal crest; in the lower lid it also has some attachment to the anterior lacrimal crest. In the lateral canthal area the septum follows and fuses with the lateral canthal tendon.

The septum is thinner medially and the entire septum weakens with age, allowing herniation of the underlying fat (Fig. 13.9). This fat is a valuable landmark to the underlying levator muscle in ptosis surgery; it can be particularly helpful in revision cases to differentiate the levator aponeurosis from the septum or surgical scarring. There are two fat pockets in the upper lid: central and nasal. These two anatomic areas are separated by some thin fascial strands from Whitnall's ligament.

Laterally in the upper lid lies the lacrimal gland, which is easily distinguished from orbital fat by location and its glandular consistency. There are three fat pockets of the lower lids: nasal, central, and temporal. The nasal fat pocket of the upper and the lower lids is whiter and more fibrous than the yellow, less dense fat of the other pockets (Fig. 13.10).

Levator Aponeurosis

The levator aponeurosis is located immediately deep to the preaponeurotic fat in the upper lid. The levator muscle arises from the lesser wing of the sphenoid in the apex of the orbit and is primarily responsible for elevation of the upper lid (Fig. 13.11). The levator muscle is innervated by the superior ramus of the third cranial nerve. The muscle thins to an aponeurosis and loses its muscle fibers in the area of Whitnall's ligament. Whitnall's ligament is a condensation of the superior sheath of the levator muscle. Whitnall's ligament attaches nasally to the fascia surrounding the trochlea and laterally to the capsule of the lacrimal gland (Fig. 13.12); it acts as a suspensory ligament of the upper lid. The muscular portion of the levator is about 40 mm long and the aponeurotic portion about 14 to 20 mm long.

Below Whitnall's ligament the levator aponeurosis fans into medial and lateral horns. The medial horn of the levator aponeurosis blends with the posterior limb of the medial canthal tendon and attaches to the posterior lacrimal crest. The lateral horn of the aponeurosis partially separates the lacrimal gland into lacrimal and palpebral portions. It inserts into the lateral canthal tendon at the lateral retinaculum. The lateral horn of the aponeurosis is considerably stronger than the medial horn. In the midportion of the upper lid the aponeurosis fuses with the orbital septum about 3 mm above the tarsus in Caucasians and a few millimeters below the tarsus in Asians (Fig. 13.2). From this point of fusion the levator aponeurosis sends fibers to attach to the orbicularis and create the lid fold. The majority of the levator aponeurosis continues inferiorly and attaches to the tarsal plate beginning about 3 mm below its superior surface.

Müller's Muscle

Immediately deep to the levator aponeurosis in the upper eyelid, originating from the terminal striated fibers of the levator muscle, lies Müller's muscle. This smooth muscle is under sympathetic control and runs about 10 mm until it attaches to the tarsus. It is closely associated with the underlying tarsus and the vascular arcade. The contraction of Müller's muscle contributes about 2 mm to the lid height.

Tarsal Plates

The tarsal plates of the upper and the lower lids are composed of dense connective tissue (not cartilage). The meibomian glands are contained within them. The tarsal plates extend from about the punctum medially to 4 to 5 mm from the lateral canthus. They become less well defined and less suitable for repairs both medially and laterally. The upper tarsus is about 10 mm in width at the midline; the lower tarsus is considerably smaller. The conjunctiva is firmly attached to the posterior tarsus in both the upper and the lower lids.

Structures of the Lower Lid

The upper and the lower eyelids have analogous structures, but they are less well defined in the lower lid (Fig. 13.13). This can be best appreciated in a cross-section of the lids and anterior orbit (Fig. 13.4). The capsulopalpebral fascia of the lower lid is analogous to the levator aponeurosis of the upper lid. It arises from the inferior rectus muscle and then splits to surround the inferior oblique muscle. When it recombines, it contributes to Lockwood's ligament and then fuses with the septum about 5 mm beneath the tarsus. It ultimately attaches to the lower border of the tarsus. The capsulopalpebral fascia, unlike the levator muscle, has no muscle fibers of its own but transmits the contraction of the inferior rectus. When an individual looks down, the transmitted motion causes the lid to depress. A dehiscence of the capsulopalpebral fascia is one cause of entropion. Deep to the capsulopalpebral fascia lies the inferior tarsal (sympathetic) muscle, which is analogous to Müller's muscle in the upper lid. Its posterior extension, like that of the capsulopalpebral fascia, surrounds the inferior oblique muscle and then attaches to the sheath of the inferior rectus to help form Lockwood's ligament (a hammock-like thickening of fascia that inserts into the lateral and the medial retinaculum). Fibrous strands between the orbital septum and the capsulopalpebral fascia separate the temporal and the central fat pockets (Fig. 13.14).

RACIAL DIFFERENCES IN EYELID ANATOMY

The major differences between the "typical" Asian and Caucasian lid are (1) Asian skin is somewhat thicker, (2) Asians commonly have a submuscular fat pocket between the orbicularis and the septum, (3) the Asian levator aponeurosis fuses with the septum below the superior surface of the tarsus (Figs. 13.15 and 13.16). These features result in a fuller lid due to the more inferior location of the preaponeurotic fat. The lid fold is lower because the levator fibers do not insert subcutaneously above the tarsus.

The eyelid crease in Asians averages 5.8 to 7.2 mm versus 7 to 10 mm for Caucasians. Similarly, the palpebral fissure is 7.6 to 9.4 mm in Asians versus 12 to 14 mm in Caucasians (Liu and Hsu, 1986). The epicanthal fold is a semilunar fold of skin extending from the upper eyelid crease to the medial canthal area. It is present in many Asians (variously reported from 40% to 90%) and occasional Caucasians. A "single eyelid" commonly means a lid with essentially no crease. A lid with a crease is often referred to as a "double eyelid." The "outer double eyelid" has a fold of 7 to 10 mm and resembles a Caucasian lid. The "inner double eyelid" has a much lower fold of 3 to 5 mm, which usually merges in the epicanthal fold (Fig. 13.2).

LACRIMAL SYSTEM

The lacrimal system is conceptually simple but physiologically complex (Fig. 13.17). The tear film is primarily produced by the lacrimal gland, with contributions from the mucin-secreting goblet cells, the accessory lacrimal exocrine glands of Krause and Wolfring, oil-producing meibomian glands, and the palpebral glands of Zeis and Moll. The tear ductules of the palpebral portion of the lacrimal gland drain into the superior lateral conjunctiva. The orbital lobe has no ductules of its own and must drain through the palpebral lobe.

The lacrimal drainage system begins with the upper and the lower puncta (Fig. 13.18). They are located 5 to 7 mm lateral to the canthus, with the inferior being slightly more lateral. The ampulla of the canaliculus is a slight dilation of its vertical portion just past the puncta before it runs horizontally into the sac. The two canaliculi usually join to form a common canaliculus that then enters the lacrimal sac superiorly and somewhat posteriorly; they may, however, enter separately. The lacrimal sac is about 15 mm in height and rests in the bony lacrimal fossa. The superior few millimeters of the sac are covered by the inferior fibers of the anterior limb of the medial canthal tendon. These fibers must usually be severed to completely expose the lacrimal sac during a dacryocystorhinostomy. The lacrimal sac empties into the nasolacrimal duct, which traverses somewhat over a centimeter of the bony canal to enter the inferior meatus in the nasal cavity.

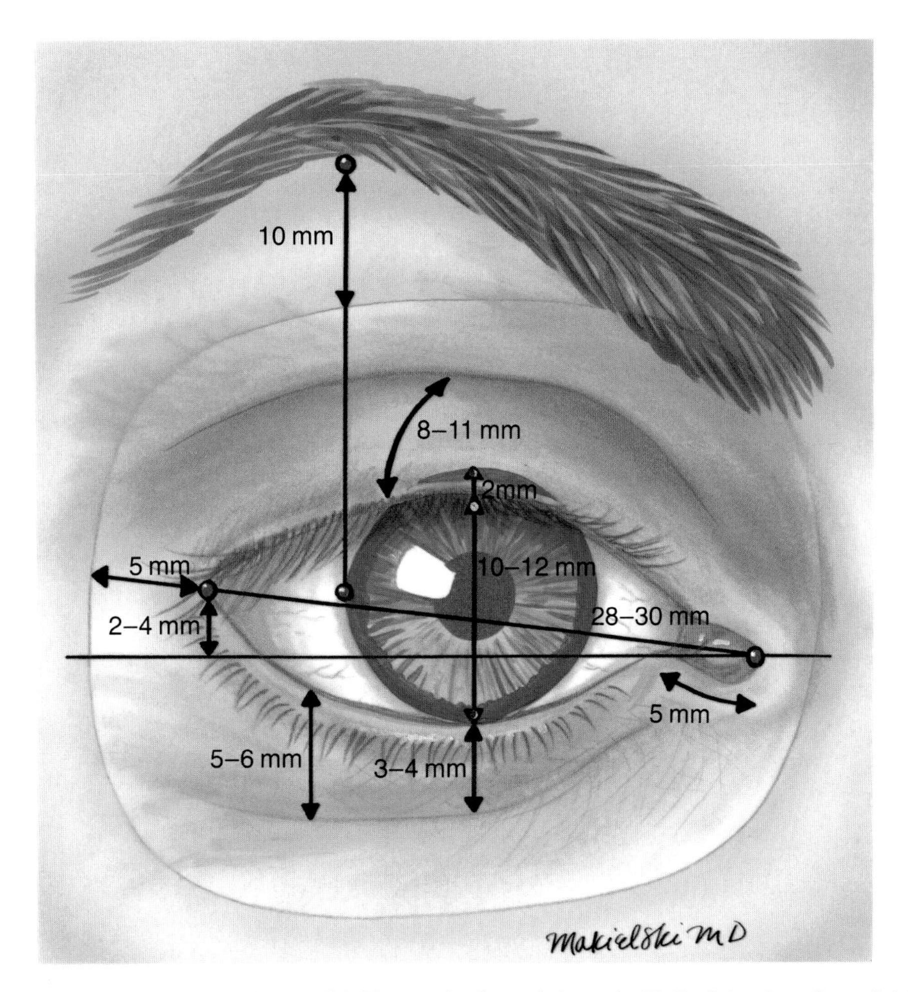

FIG. 13.1. Topography of the eyelid. The eye is almond shaped with the lateral canthus slightly more superior than the medial canthus; typical superior elevations at the lateral canthus are 2 mm for men and 4 mm for women. An average palpebral opening is 10 to 12 mm in height and 28 to 30 mm in width. The distance from the lateral canthus to the orbital rim is about 5 mm. The upper lid fold in Caucasians is approximately 8 to 11 mm. The lower lid crease is about 5 to 6 mm. The high point of the brow is directly superior to the lateral limbus. The distance from the supraorbital rim to the inferior aspect of the brow at the lateral limbus is about 10 mm in Caucasian women.

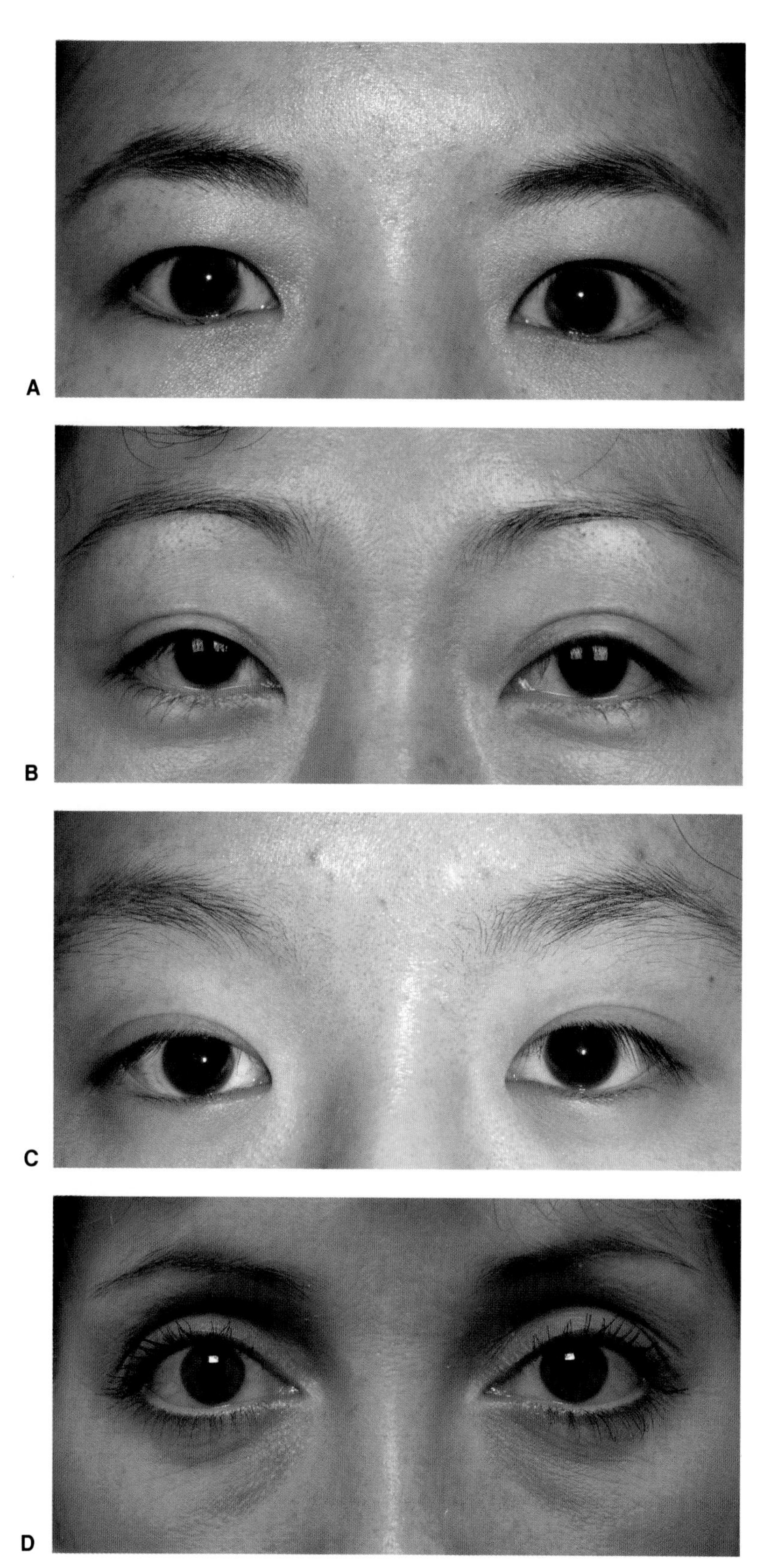

FIG. 13.2. Racial variations in upper eyelid topography. A: An upper lid without a crease is termed a single lid. An epicanthal fold is present. **B:** An outer double eyelid with a crease 7 to 10 mm above the lashes. The epicanthal fold is not noticeable. **C:** An inner double lid with a crease 3 to 5 mm above the lashes and a more prominent epicanthal fold. **D:** A non-Asian eyelid with a crease 8 to 10 mm above the lashes and no epicanthal fold.

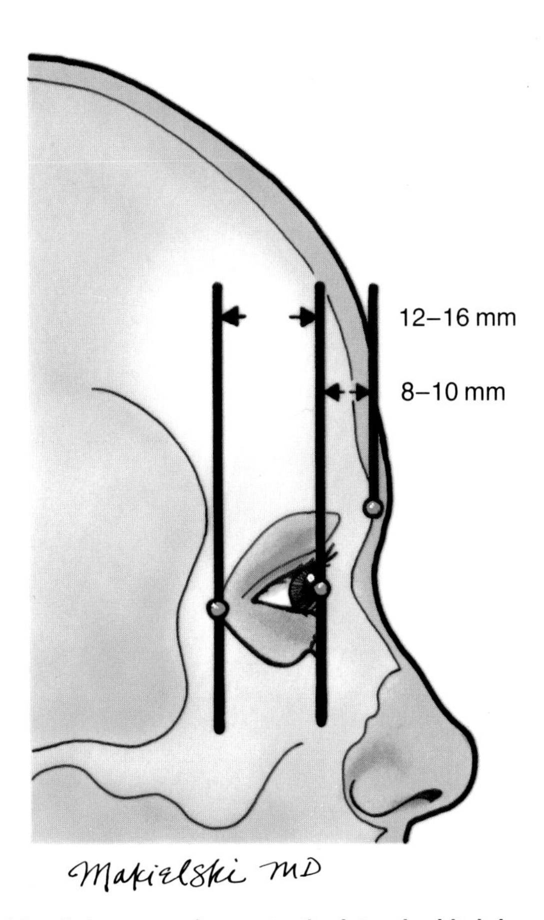

FIG. 13.3. Relationship of the corneal apex to the lateral orbital rim and the supraorbital ridge. The anterior-posterior distance from the lateral orbital rim to the apex of the cornea is typically 12 to 16 mm and the distance from the supraorbital ridge to the cornea is 8 to 10 mm. The most commonly used of these two measurements is the distance from the lateral orbital rim to the corneal apex as measured with the Hertel exophthalmometer. The relative position of the orbital bones and the size of the globe both affect the result, therefore no absolute interpretation can be placed on a given number. The measurements are most useful in comparing the two eyes (e.g., in a posttraumatic enophthalmos) or in following a patient with the progressive exophthalmos of Graves' disease.

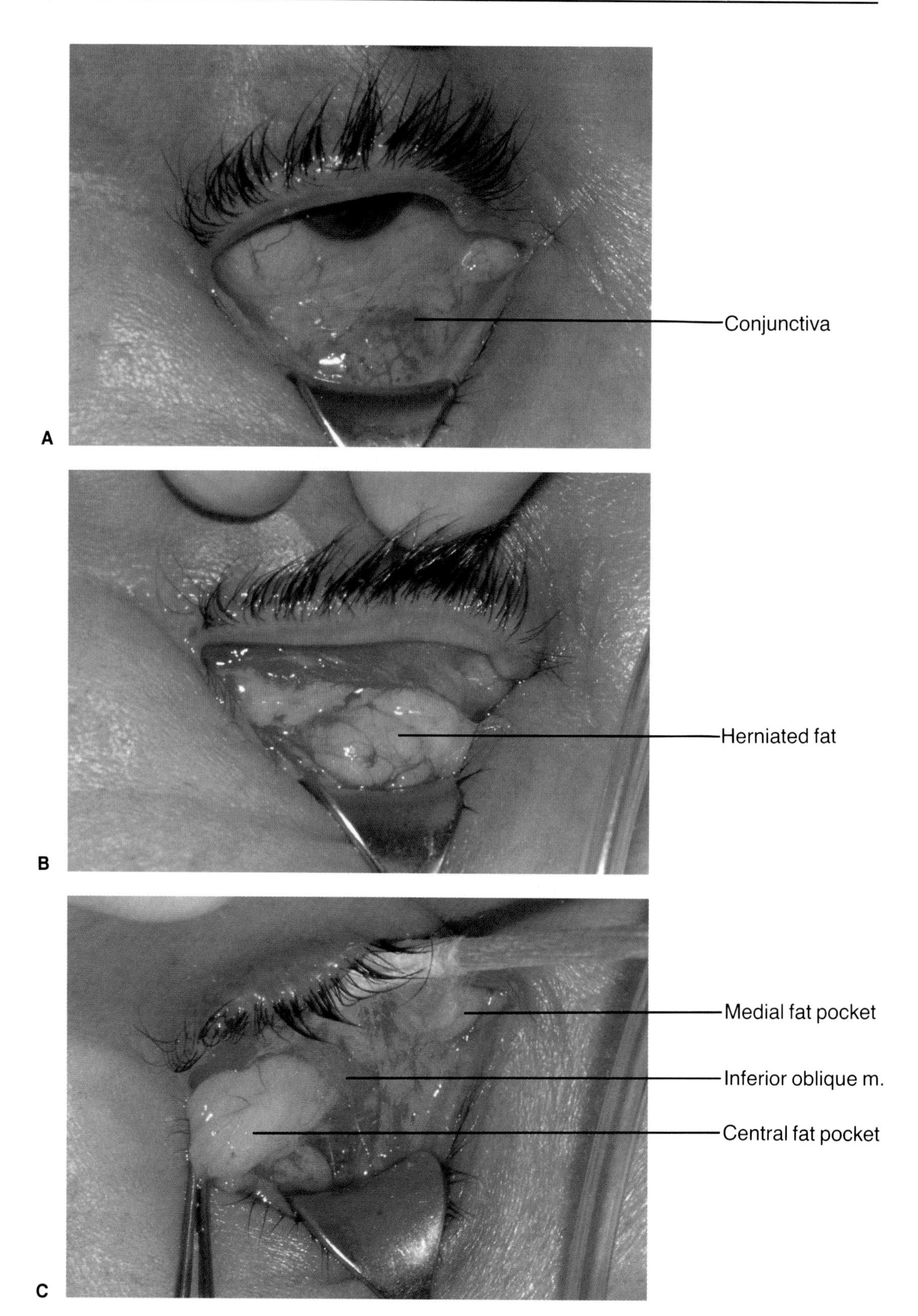

FIG. 13.4. Lower lid from the transconjunctival approach. A: The conjunctiva is exposed. **B:** The conjunctiva and the retractors incised; herniated fat is visible. **C:** The inferior oblique muscle runs between the medial and the central fat pockets and, along with the orbital rim, is a landmark for the depth of fat excision.

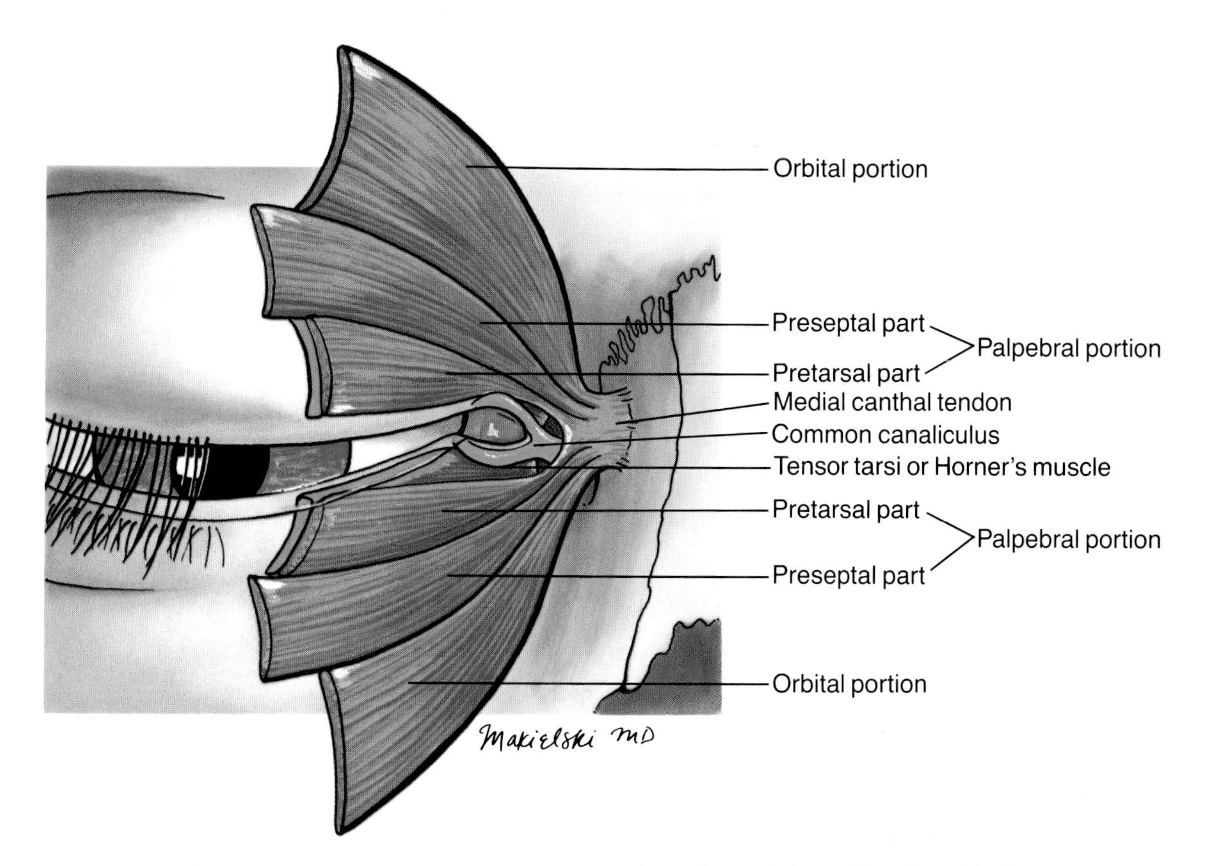

FIG. 13.5. Orbicularis oculi muscle. The orbicularis oculi muscle is traditionally divided into an orbital portion and a palpebral portion. The palpebral portion is subdivided into a preseptal and a pretarsal part. Both the preseptal and the pretarsal muscles have a deep head, which arises from the posterior lacrimal crest and fossa, and a superficial head, which arises from the medial canthal tendon. These two limbs surround the lacrimal sac and create a pumping mechanism. The deep head of the pretarsal part is termed the tensor tarsi or Horner's muscle.

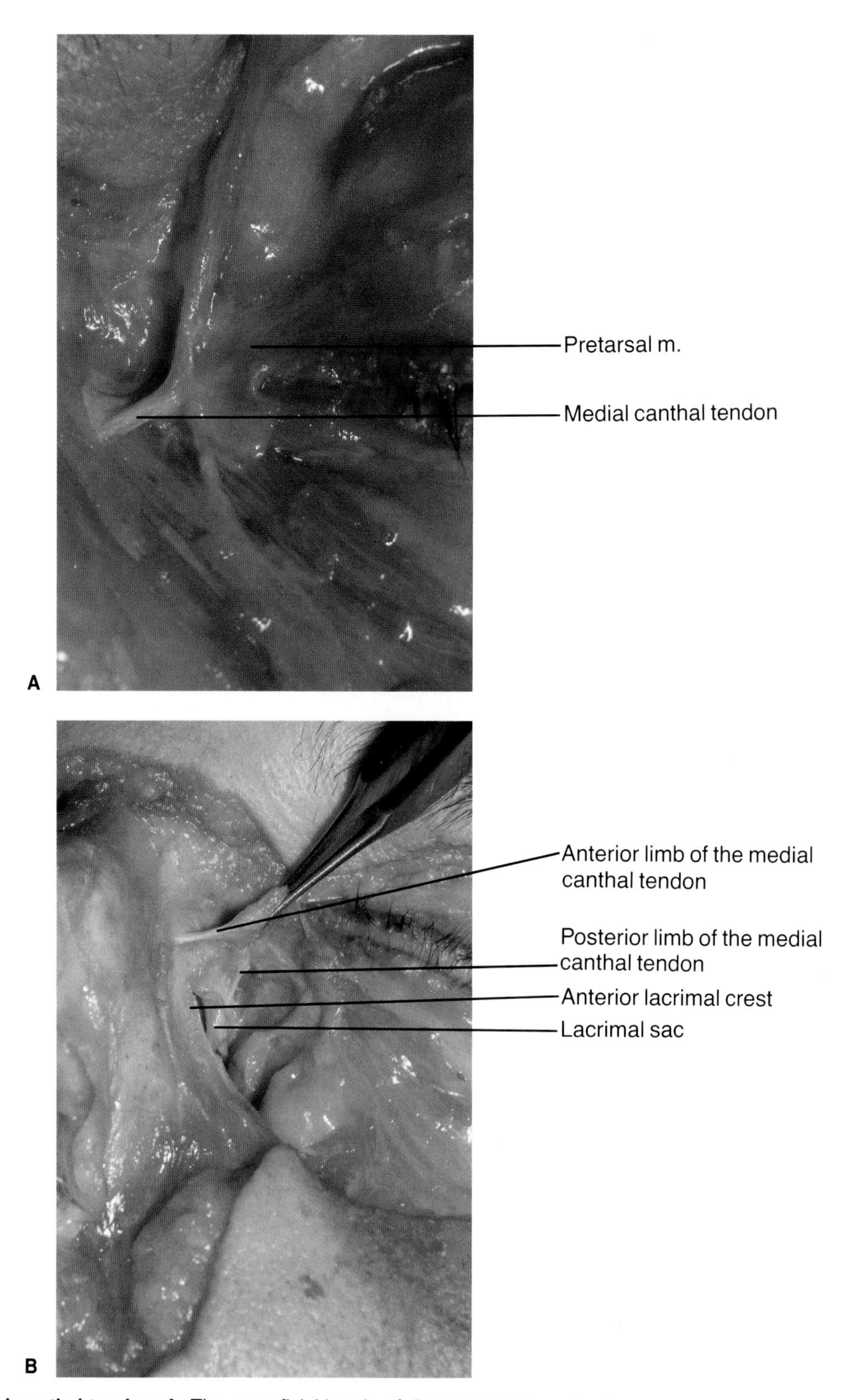

FIG. 13.6. Medial canthal tendon. A: The superficial heads of the pretarsal muscles become the medial canthal tendon. **B:** The anterior limb of the medial canthal tendon is reflected superiorly to demonstrate the posterior limb of the medial canthal tendon. The superior lacrimal sac lies between these two limbs and the inferior aspect of the medial canthal tendon must be incised to completely expose it. The anterior lacrimal crest marks the anterior boundary of the lacrimal fossa.

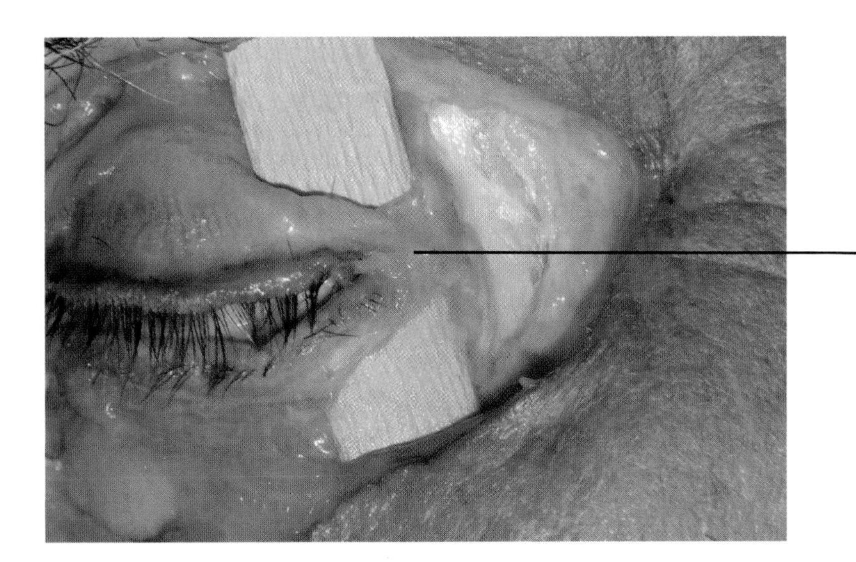

FIG. 13.7. Lateral canthal tendon. The lateral canthal tendon inserts into the orbital tubercle about 5 mm behind the rim.

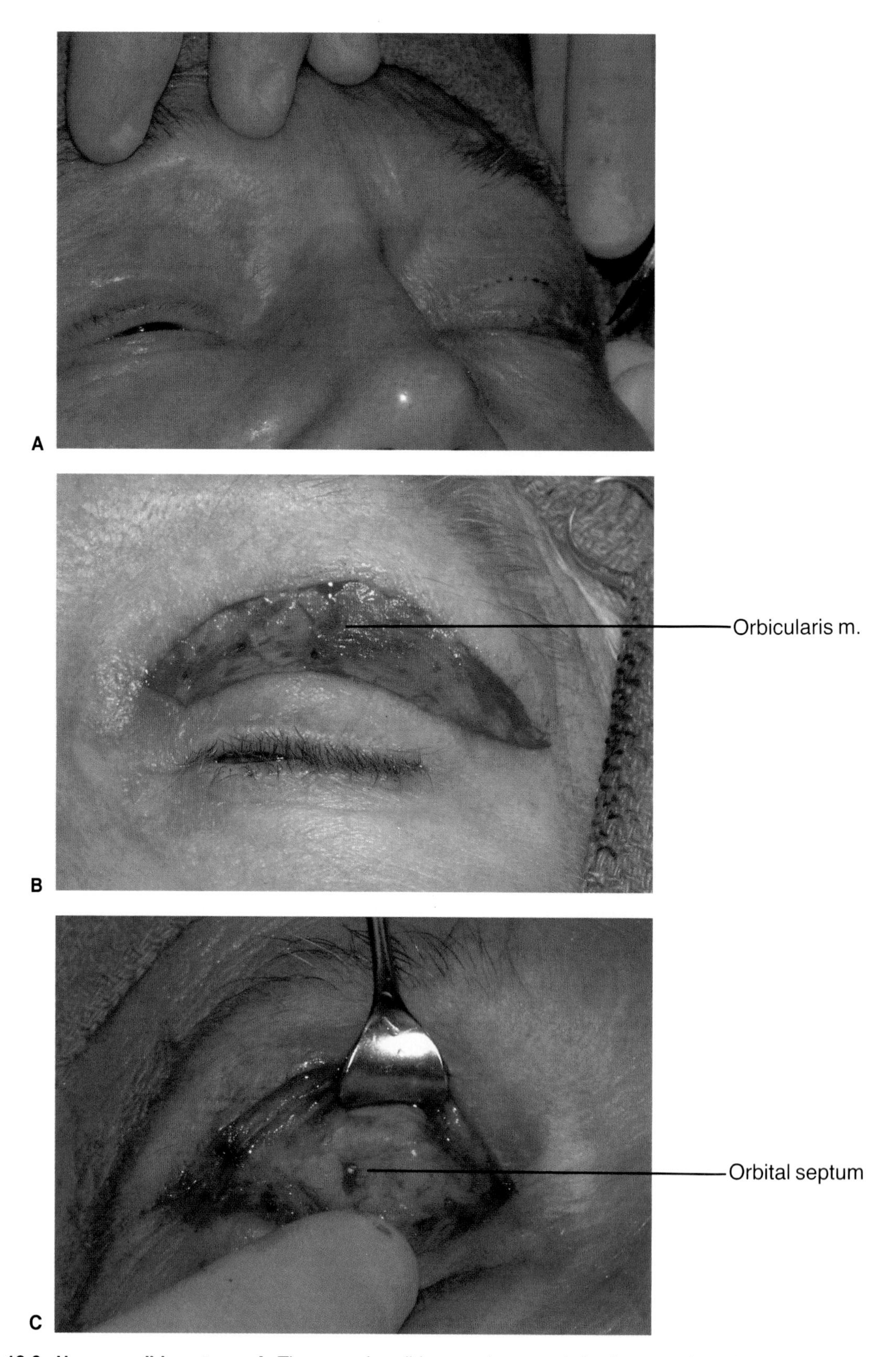

FIG. 13.8. Upper eyelid anatomy. A: The normal eyelid crease in a non-Asian is approximately 8 to 10 mm above the lash line. **B:** The orbicularis muscle lies immediately beneath the skin. **C:** The orbital septum bulges with slight digital pressure.

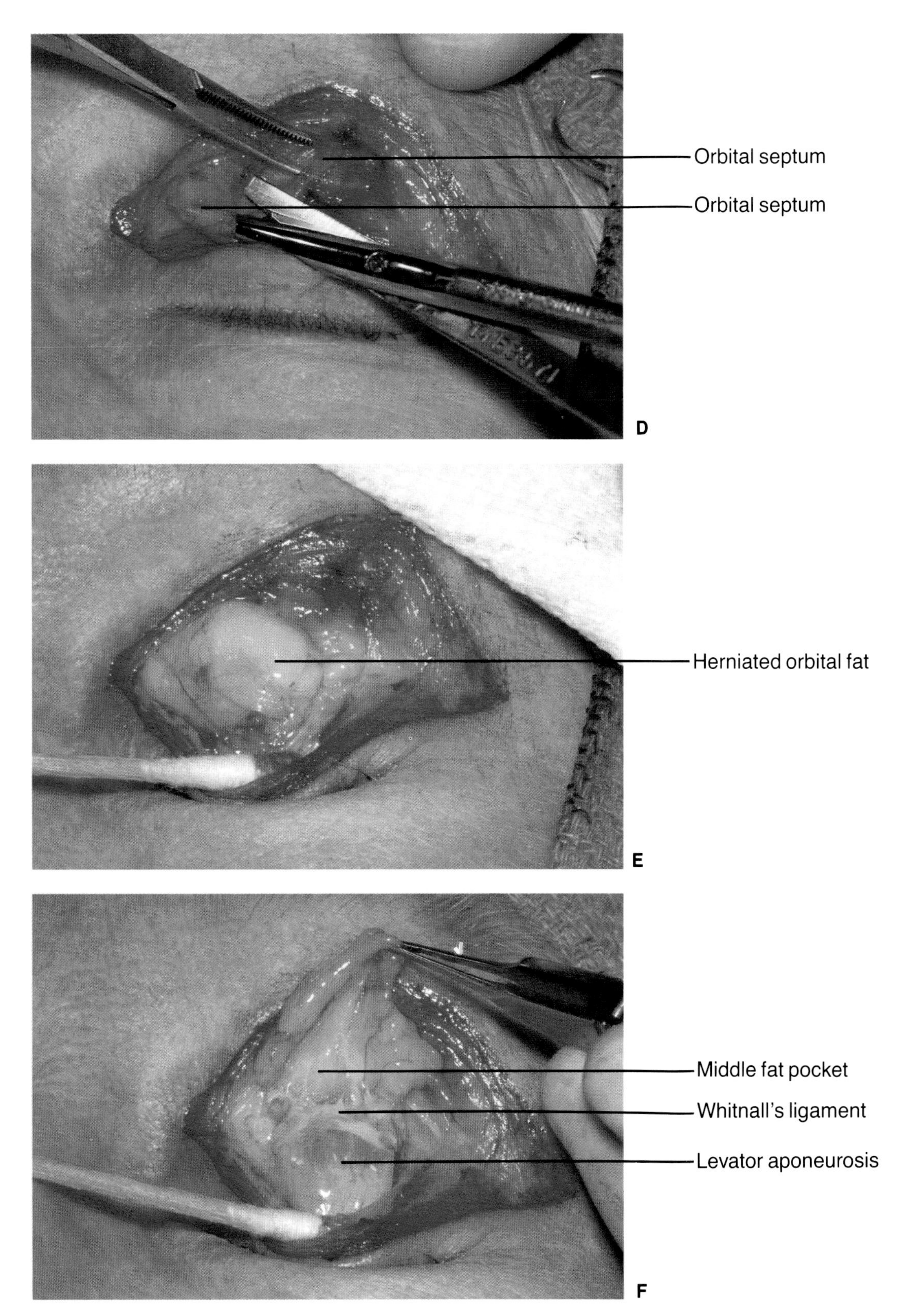

FIG. 13.8. *Continued.* **D:** The orbital septum is opened and demonstrated over instruments. **E:** Herniated orbital fat. **F:** Fat is retracted to demonstrate the deeper levator aponeurosis with Whitnall's ligament.

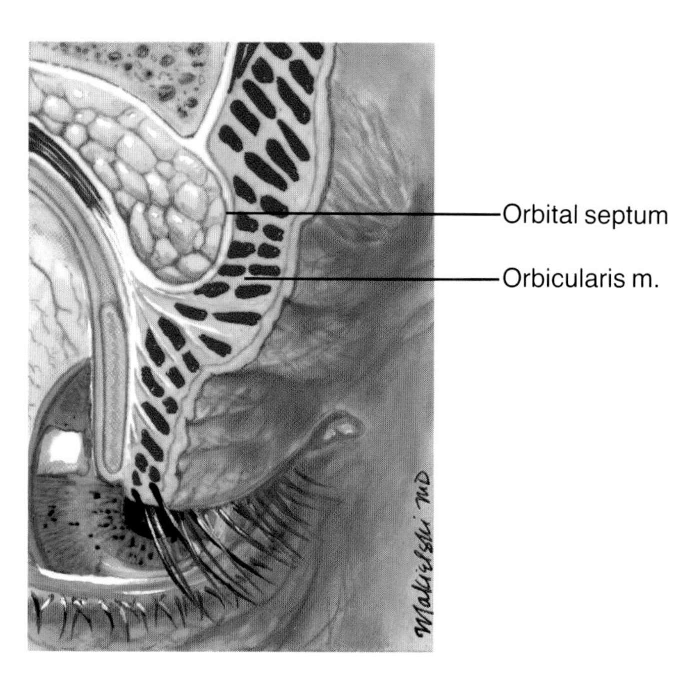

FIG. 13.9. Aging upper eyelid. The orbital septum and the orbicularis muscle are weakened with age, resulting in herniation of orbital fat.

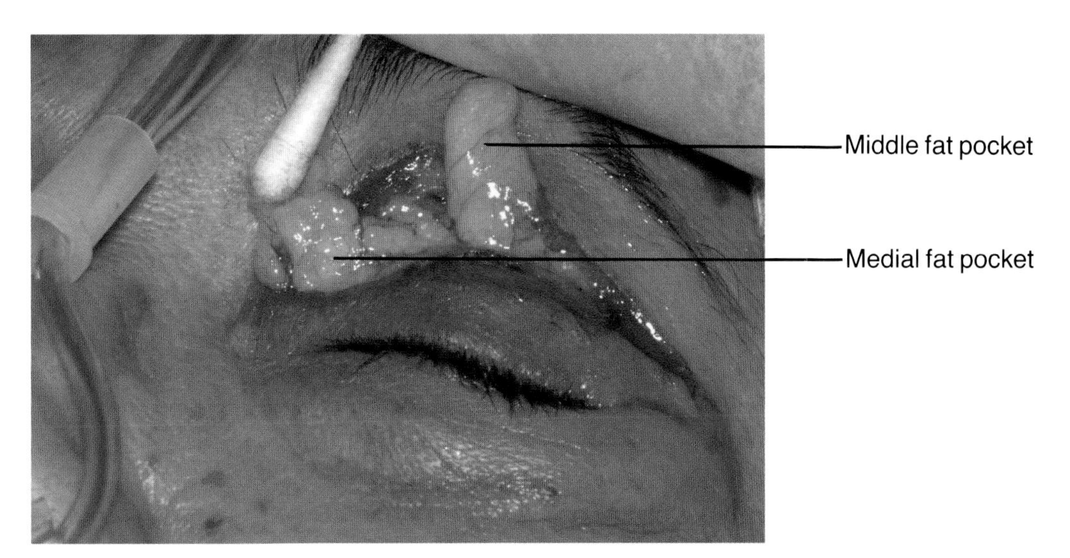

FIG. 13.10. Eyelid fat pockets. The medial fat pockets in both the upper and the lower eyelids are whiter and more fibrous than the central fat pockets.

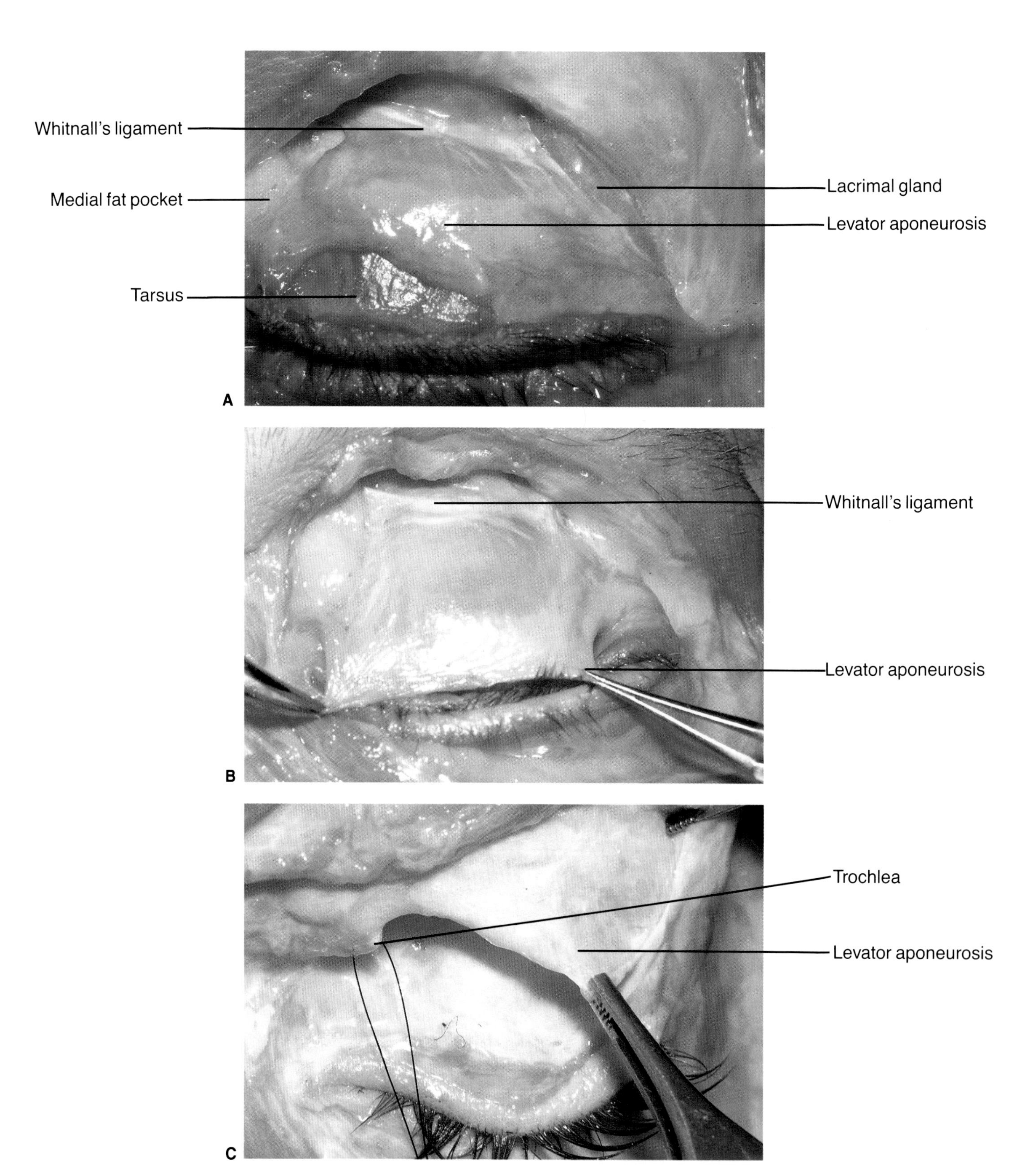

FIG. 13.11. Levator aponeurosis. A: The levator aponeurosis with the medial fat pocket and the lacrimal gland. **B:** The levator extended to better demonstrate Whitnall's ligament. **C:** Whitnall's ligament is attached to fascia around the trochlea medially.

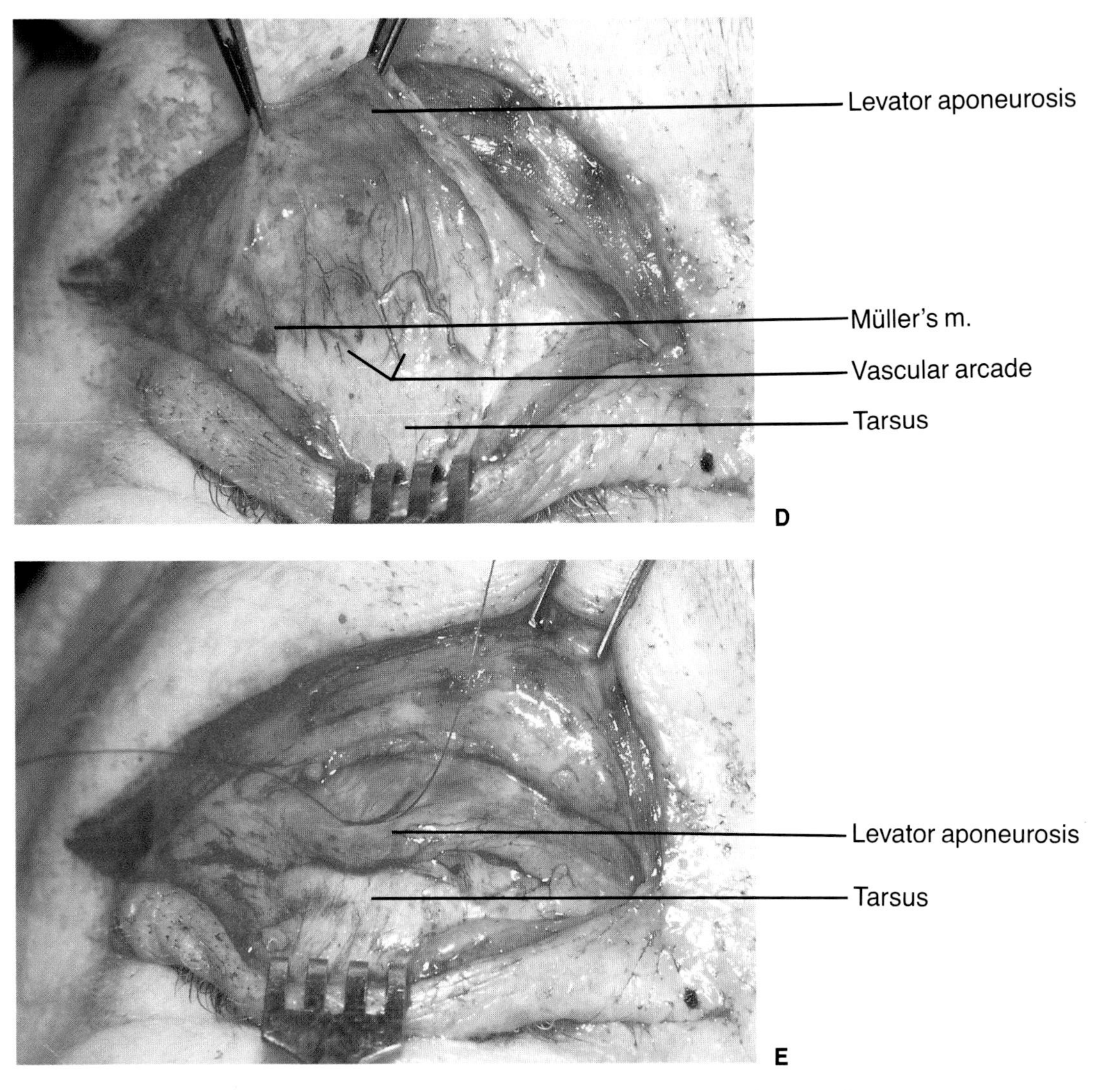

FIG. 13.11. *Continued.* **D:** The levator aponeurosis elevated, showing the tarsus, the vascular arcade, and Müller's muscle. **E:** The levator aponeurosis is advanced and sutured to the tarsus in a ptosis repair.

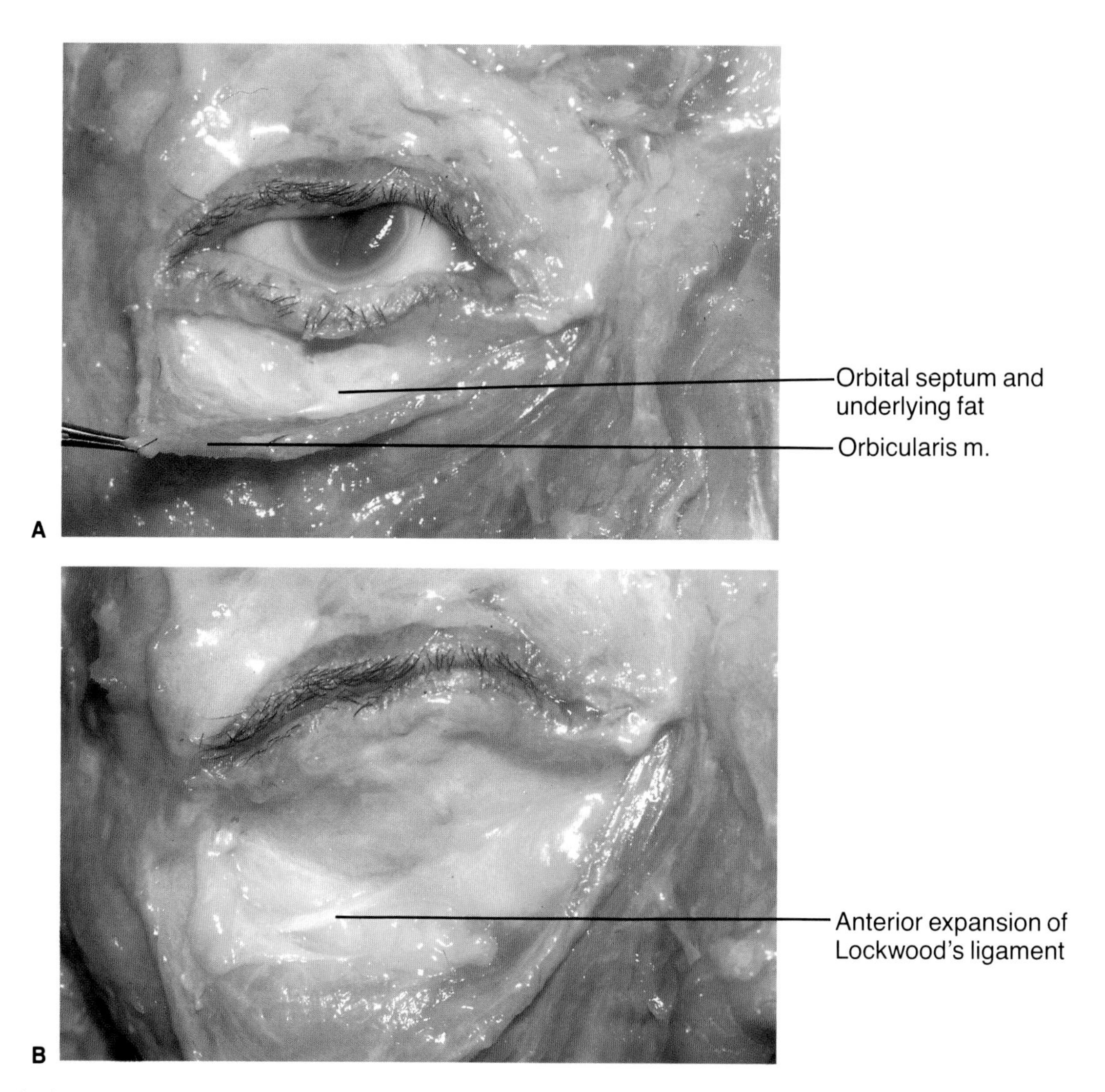

FIG. 13.12. Lower eyelid anatomy. A: The orbicularis muscle is retracted to reveal the orbital septum and the underlying fat. **B:** With the septum opened, the anterior expansion of Lockwood's ligament (also called the arcuate expansion of the inferior oblique) is seen to separate the central from the lateral fat pocket.

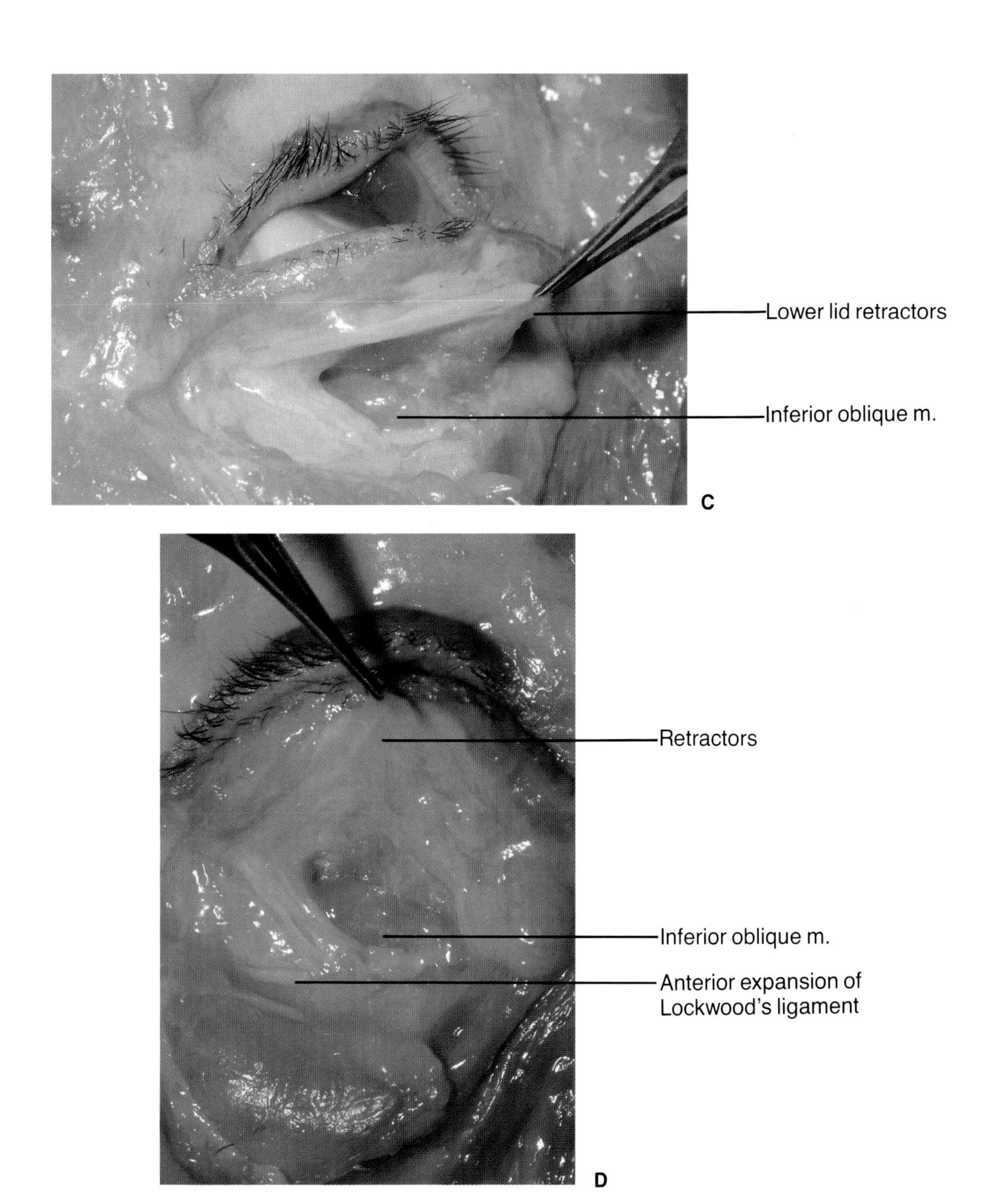

FIG. 13.12. *Continued.* **C:** The lower lid retractors are elevated to demonstrate the inferior oblique muscle. The inferior oblique muscle separates the medial fat compartment from the central fat compartment. **D:** The inferior oblique muscle, the retractors, and the anterior expansion of Lockwood's ligament are seen.

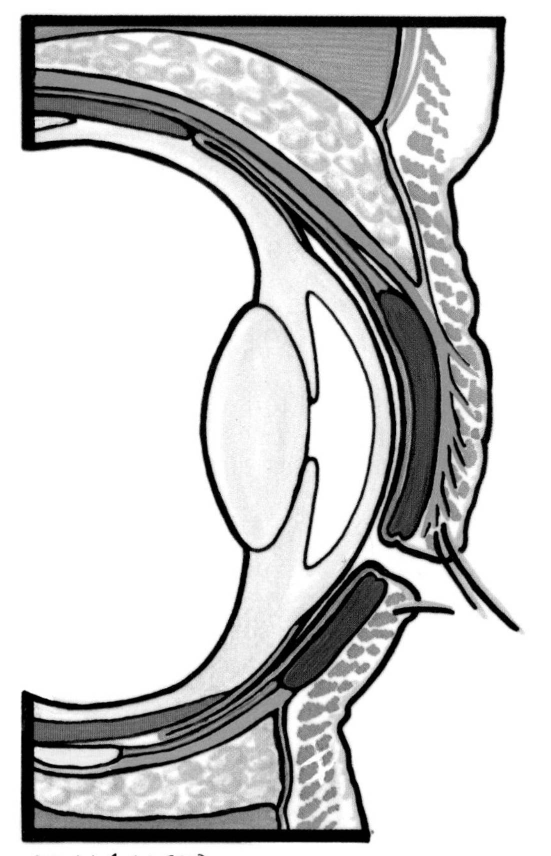

FIG. 13.13. Cross-section of the eyelids: paired structures. The upper and the lower eyelids have analogous structures with similar functions. The paired tarsal plates are seen in purple. The levator muscle, which primarily elevates the upper lid, is analogous to the capsulopalpebral fascia of the lower lid; both are seen in green. Unlike the levator, the capsulopalpebral fascia has no muscular fibers but transmits the actions of the inferior rectus. Deep to these structures are the sympathetically innervated Müller's muscle and its lower lid counterpart, the inferior tarsal muscle, seen in blue. The inferior and the superior rectus muscles are seen in orange. The superior oblique tendon and the inferior oblique muscle are designated in yellow.

A

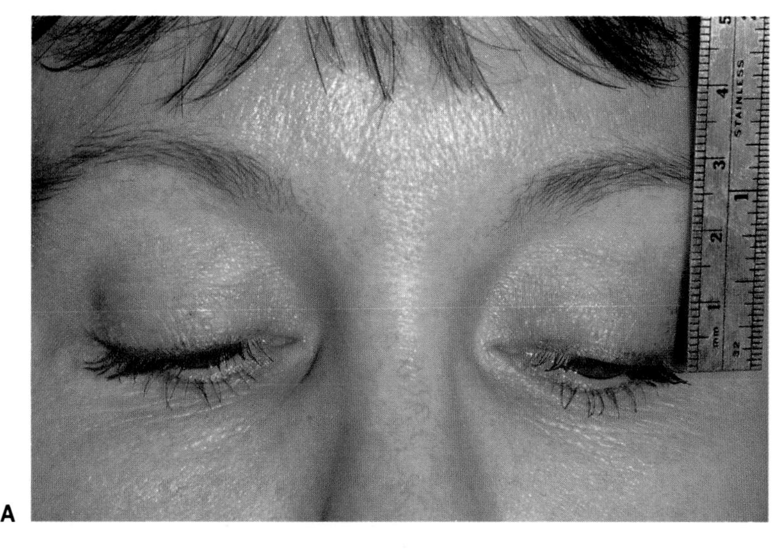

B

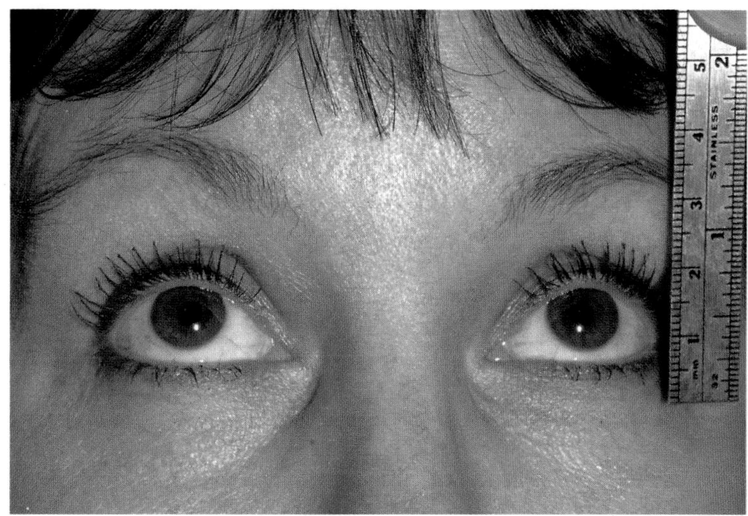

FIG. 13.14. Function of the levator muscle. A, B: Full lid excursion is measured by comparing the upper lid position while looking down and looking up. The brow should be manually stabilized to prevent motion of the frontalis muscle; 15 to 18 mm of excursion is normal. Levator function is typically fair to good with acquired ptosis and is often poor with congenital ptosis.

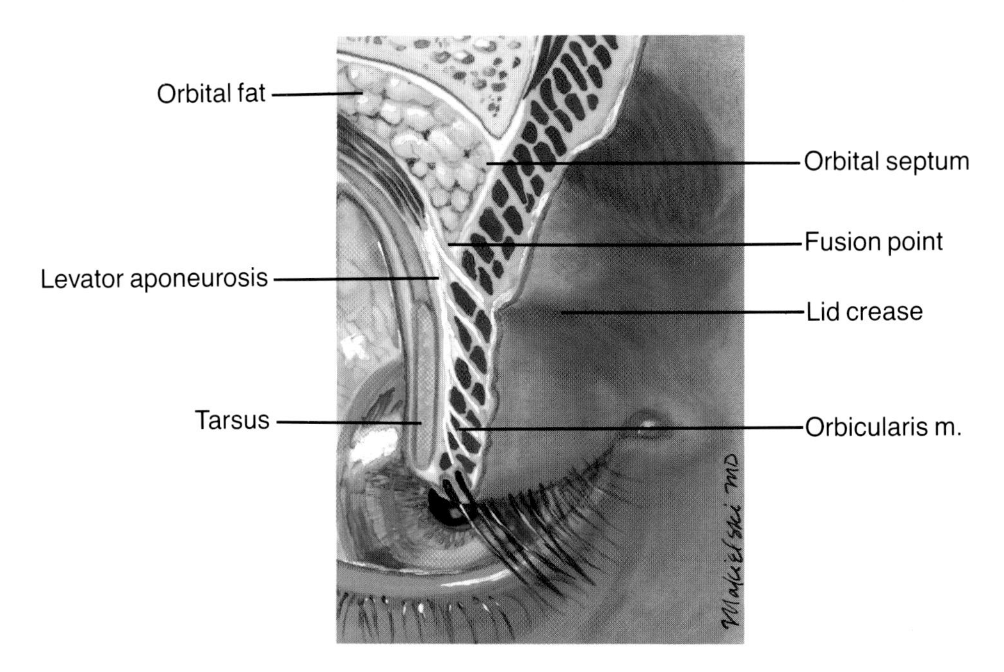

FIG. 13.15. Caucasian/African (non-Asian) upper eyelid. The orbital septum and the levator aponeurosis fuse above the tarsus. Fibers from the levator aponeurosis extend to the dermis to create the lid fold.

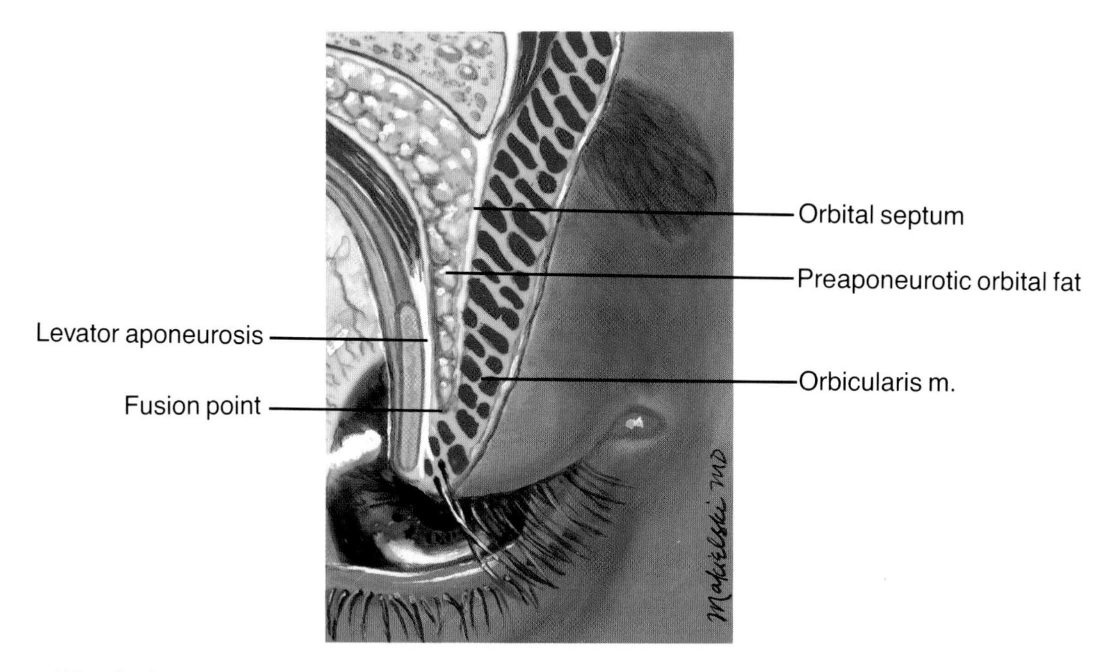

FIG. 13.16. Asian upper eyelid. The orbital septum and the levator aponeurosis fuse below the superior edge of the tarsus. Preaponeurotic orbital fat extends inferiorly between the levator aponeurosis and the orbicularis muscle.

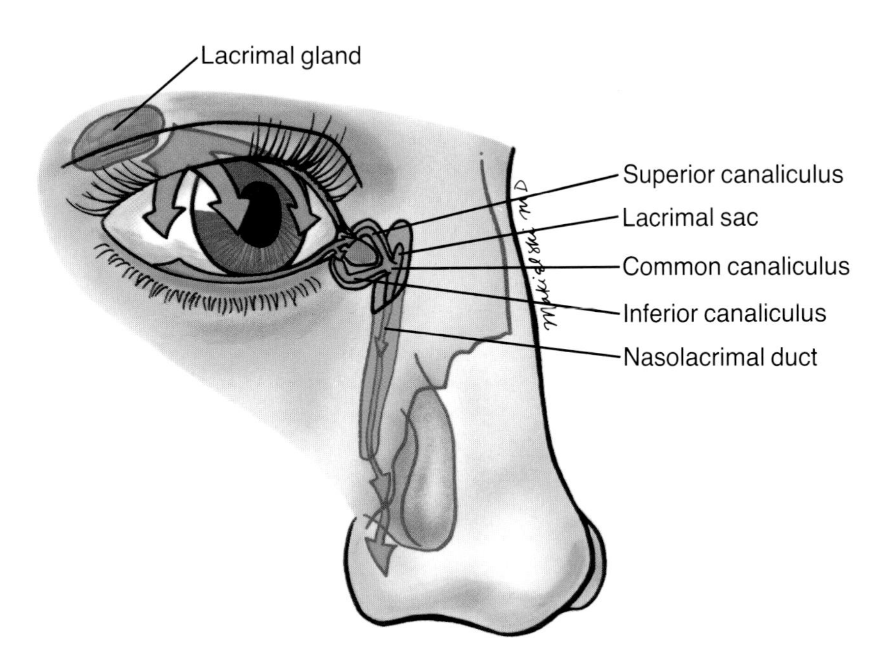

FIG. 13.17. Lacrimal system. The tear film produced by the lacrimal gland and the multiple accessory gland reaches the upper and lower puncta located 5 to 7 mm from the medial canthus. The lower puncta is usually slightly more lateral than the medial canthus and is the more important of the two. The puncta opening is about 0.3 mm in diameter but can be easily dilated to probe the ducts. Immediately beneath the puncta lies the ampulla, an enlargement of the canaliculi before they turn horizontally. The two canaliculi join to form a common duct prior to draining into the lacrimal sac. The lacrimal sac is about 15 mm in height and empties into the nasolacrimal duct, which is about 10 to 12 mm in length. It drains into the inferior meatus about 15 mm from the nasal floor. Intranasally, the lacrimal sac is located at the level of the middle meatus.

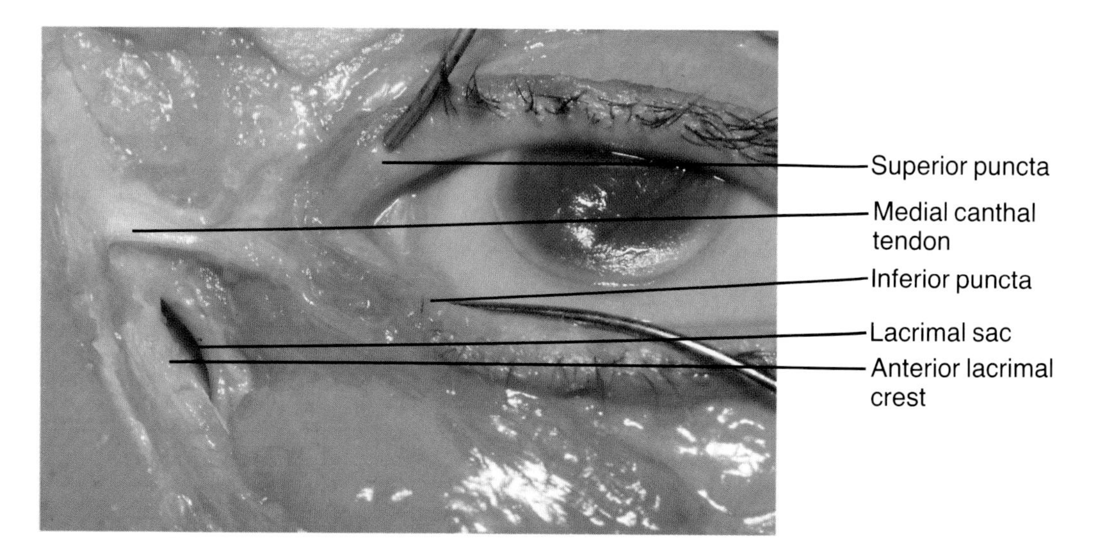

FIG. 13.18. Lacrimal sac and fossa. Probes enter the superior and the inferior puncta to outline the canaliculi.

CHAPTER 14

Nose

The complex contours of the nose, more than any other anatomic feature, define and characterize the face. The surface anatomy reflects a combination of the structure of the underlying bony/cartilaginous framework and the draping of the skin and the soft tissue. Although the contour cannot be analyzed very simply, for reconstructive purposes it can be visualized as a combination of discrete planes that define aesthetic subunits: the dorsum, the sidewall, the soft-tissue triangle, the ala, and the tip (Fig. 14.1). Whenever possible one should place incisions at the junctions of these planes and honor the integrity of the individual aesthetic subunit.

AESTHETIC NASAL PROPORTIONS

From a frontal view the nose should provide a symmetric confluence of the various aesthetic subunits. Important landmarks include the glabella, the nasion, the rhinion, the supertip, the tip-defining points, the ala, the alar cheek junction, the lobule, the alar lobular junction, the supra-alar crease, the columella, and the external soft-tissue triangle (Figs. 14.2, 14.3, and 14.4). There should be a smooth curve from the brows to the nasal dorsum (Fig. 14.5). The nasal tip demonstrates a light reflex from the two tip-defining points.

Several simple proportions relate the nose to the rest of the face. The width of the nasal base is approximately the intraocular distance. The nose represents approximately one-third of the distance from the hairline to the chin. The nasion or nasofrontal angle is approximately at the level of the upper lid folds. These same tomographic points are important in the lateral, oblique, and basal views. Nasal projection is best appreciated from a lateral view. It is most simply measured using a 3-4-5 triangle as described by Crumley and Lancer (Fig. 14.6). From the lateral view one also sees a gentle supertip break and ideally 2 or 3 mm of columella. The columella should have a gentle double break (a straight columella due to inappropriate shortening of the septum is an unattractive surgical stigmata) (Fig. 14.7).

The ala and the columella should be analyzed both separately and as a complex. The columella and the alar margins should maintain an angle to the Frankfort horizontal of about 20° or to the facial plane of about 70°. Subtle irregularities such as the hanging columella and hooding of the ala or variations in their inclinations to the facial plane need to be addressed specifically in rhinoplasty procedures (Fig. 14.8).

In general, one controls the columella by changing the length of the septum and the medial crura (shortening with excision or extending with grafts) and changes the ala by excision or grafting of the lateral crura and direct soft-tissue excision. The angle between the lip and the columella should be neither too acute nor too blunted; 90° to 115° is an aesthetic range, with men having a more acute angle.

The oblique view is mandatory for study of the profile (Fig. 14.9). The gentle curve seen from the supraorbital rim down to the dorsum is accentuated, and one sees the relationship of the lobule, the columella, and the ala. The basal view allows evaluation of the nostril shape and the relationship between the columella and the lobule. The nostril should be somewhat pear-shaped and the ratio of the columella length to the lobule height about 2:1 (Fig. 14.10). A somewhat larger lobule (approaching 1:1) can be pleasing. An extremely small lobule (usually with elongated nostrils) is generally unattractive. Care should be taken not to accentuate this disparity when reducing the overprojecting tip with this type of nose. The basal view also shows any caudal septal deflections and variations in the medial crura (e.g., splaying of the feet). Longstanding asymmetries of the nostril shape are fairly common and should be brought to the patient's attention preoperatively; they are usually difficult to correct. A patient in whom a significant decrease in nasal projection is planned will have a corresponding change in the basal relationships, including rounding of the nostrils and possible alar flaring. These patients may require alar base narrowing at the time of their primary surgery or as a secondary procedure.

SOFT-TISSUE CONSIDERATIONS

Although most effort in rhinoplasty is addressed to the bony and cartilaginous framework, the skin and associated soft tissue are equally important. Skin thickness is a major variable in rhinoplasty and one over which the surgeon has little control. Thick-skinned individuals will not obtain a sculpted appearance in rhinoplasty despite the surgeon's best efforts. Extremely thin-skinned individuals will reveal even very minor irregularities and asymmetries. These limitations should be discussed with the patient. The surgeon should be aware of these factors when selecting the best technique; for example, vertical dome division may be more appropriate in a thick-skinned than a thin-skinned person.

Skin thickness varies significantly over the nose (Fig. 14.11). The skin is thicker over the nasion, thinner over the rhinion, and thicker again over the supertip. When lowering the bony and cartilaginous dorsum, the surgeon must preserve a natural prominence in the area of the rhinion to conform to the thickness of the overlying skin and the soft tissue.

In the nose, the SMAS is a distinct but delicate meshwork overlying the nasal framework and analogous to the SMAS of the remainder of the lower face (Fig. 14.12). The skin and the soft tissue envelope (fat, nasal mimetic musculature, fascia) should normally be elevated in one layer and preserved. All dissection should be performed in the plane immediately above the perichondrium and deep to the periosteum. This technique preserves a layer of soft tissue to create subtle surface contours and to camouflage minor bony and/or cartilaginous irregularities. It is also a reasonably avascular plane. Nasal lymphatics follow the veins; dissection in the midline and deep to the SMAS preserves those lymphatics and minimizes edema (Fig. 14.13).

The major nasal muscles of clinical importance are the depressors. In a nose with poor tip support, the depressors contribute to the "plunging tip" with facial motion. Transection of these muscles can be an important adjunct to preserving tip support and projection in these patients. The procerus can be a significant element in the nasofrontal angle and can be cut transversely to deepen it.

A majority of the ala is soft tissue, with the cartilage of the lateral crura running superiorly (Fig. 14.14). Alar reduction must therefore be created primarily with soft-tissue excision. The relationship between the alar margin and the lateral crura is important when making a cartilage-splitting incision; an incision that parallels the alar margin will cut across the lateral crura and create undesired rotation.

HARD-TISSUE SUPPORT

The septum has an important role in nasal physiology as well as nasal contour. The caudal septum, particularly important in septal surgery, has three angles that must be preserved in surgery (Fig. 14.15). The septum provides support and shape to the dorsum between the nasal tip and the nasal bones. Strong dorsal and caudal struts will, in most cases, preserve the profile following nasal surgery. The nasal spine provides essential support to the caudal septum; the shape of the spine can contribute to caudal deviations and blunting of the lip and the columellar angle. The nasal valve controls the flow of air into the nasal cavities. It is a teardrop-shaped area with the tip at the angle between the septum and the upper lateral cartilage, and the base bounded by the premaxilla, the piriform aperture, and the anterior end of the inferior turbinate.

The complex contours of the lower lateral cartilage largely define the nasal tip shape and support. The tip-defining point is generally at the most medial edge of the lateral crura. The entire lower lateral cartilage is quite variable both among individuals and from side to side in the same individual. The feet of the crura attach to the caudal septum. The intermediate crura represents a sometimes poorly defined area between the lateral and the medial crura. There is frequently a dramatic narrowing of the lower lateral cartilage just medial to the tip-defining areas. The lateral crura itself is usually slightly convex but occasionally has a concavity at the hinge point.

Major tip support mechanisms include the interdomal sling, attachment of the medial crura to the septum, attachment of the lower lateral cartilage to the upper lateral cartilage, and the intrinsic strength of the lower lateral cartilage with attachments to the piriform aperture (Figs. 14.16, 14.17, and 14.18).

The tripod concept explains tip dynamics and the effects of various surgical procedures on tip rotation and projection. The conjoint medial crura and two lateral crura form the legs of a tripod (Fig. 14.19). By changing the pivot point of the legs, the tip can be rotated. Shortening the legs of the tripod can decrease projection as well as cause rotation. The surgeon, by excising across the lateral crura, can change the pivot point and obtain rotation or remove a section of the lateral crura and also drop projection. Similarly, one can decrease projection by excising a portion of the medial crura.

The upper lateral cartilage plays a major role in nasal physiology and also creates the nasal sidewall and the middle third of the nose. The upper lateral cartilage is attached to the septum and the undersurface of nasal bones and frequently interdigitates with the lower lateral cartilage in the scroll area (Figs. 14.20 and 14.21).

The paired nasal bones articulate with the frontal process of the maxilla and the frontal bone to create the bony vault. The length of the nasal bones is quite variable; patients may have quite short nasal bones that contribute little to the profile. Nasal bones are usually slightly convex but can be concave (usually secondary to trauma) or markedly convex. The size and the shape of the nasal bones determine the location and the type of osteotomy performed in rhinoplasty surgery. There is a visual relationship between the height and the width of the nose, with a higher dorsum appearing narrower from the frontal view and vice versa. The nasal bones become thicker as they progress more cephalad (Fig. 14.22).

Standard lateral and medial osteotomies create a fracture that extends from the nasofacial junction to the medial canthus and toward the nasion where it meets the small back fracture of the medial osteotomy. Osteotomy is not feasible (and is aesthetically undesirable) through the thick cephalic nasal bone. The superior osteotomy site should approximate the junction of this thick and thinner bone (Fig. 14.23). In thick noses, debris and bone must be removed cephalically to allow adequate narrowing. In wide or extremely convex or asymmetric nasal bones, intermediate osteotomies are made between medial and lateral osteomies to narrow and equalize the length of the sides.

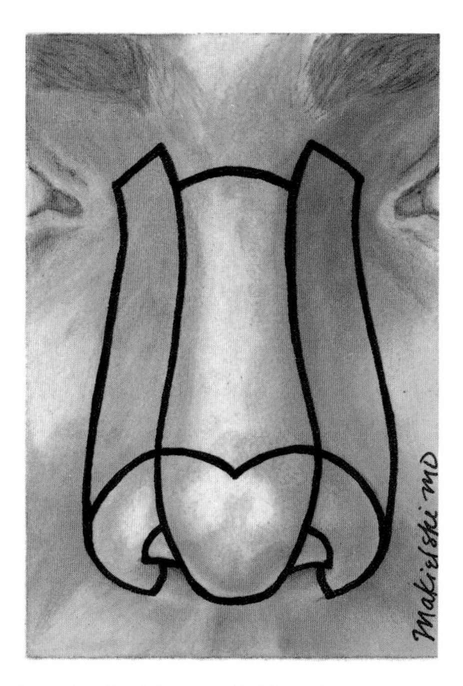

FIG. 14.1. Nasal aesthetic subunits. The nasal sidewalls, the dorsum, the ala, the tip, and the soft-tissue triangles constitute discrete subunits with similar contour characteristics. Reconstruction of an entire subunit is desirable.

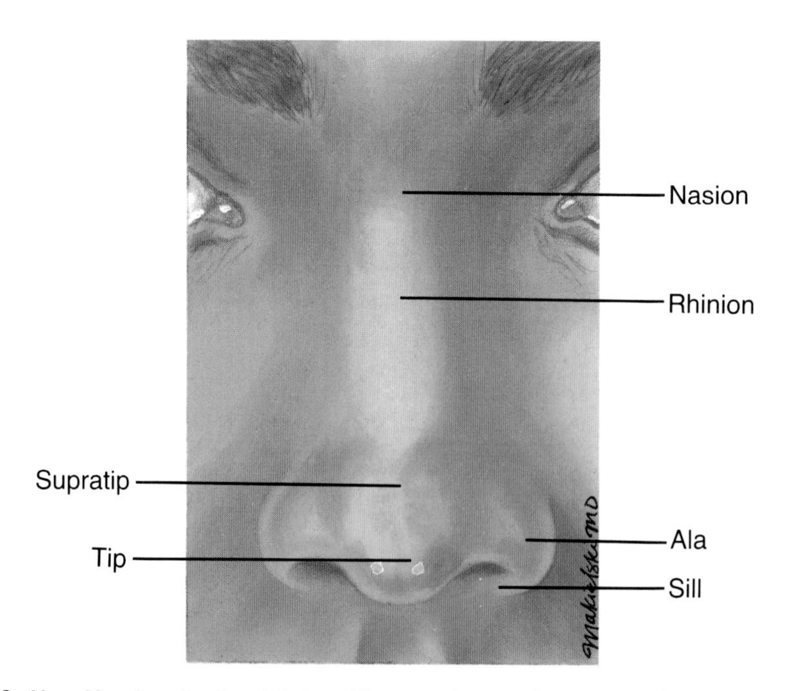

FIG. 14.2. Nasal landmarks: frontal view. The complex nasal contour and its landmarks represent a combination of the bony and the cartilaginous framework and the overlying skin–soft-tissue envelope. For example, the nasion corresponds to the position of the nasofrontal suture, but its depth is also affected by the procerus muscle. The perceived width of the nose in this view is greatly affected by the height of the nasal dorsum—a higher dorsum creates a narrower appearing nose. Normal lower lateral cartilage shows a symmetrical double light reflex at the dome-defining point in the frontal view.

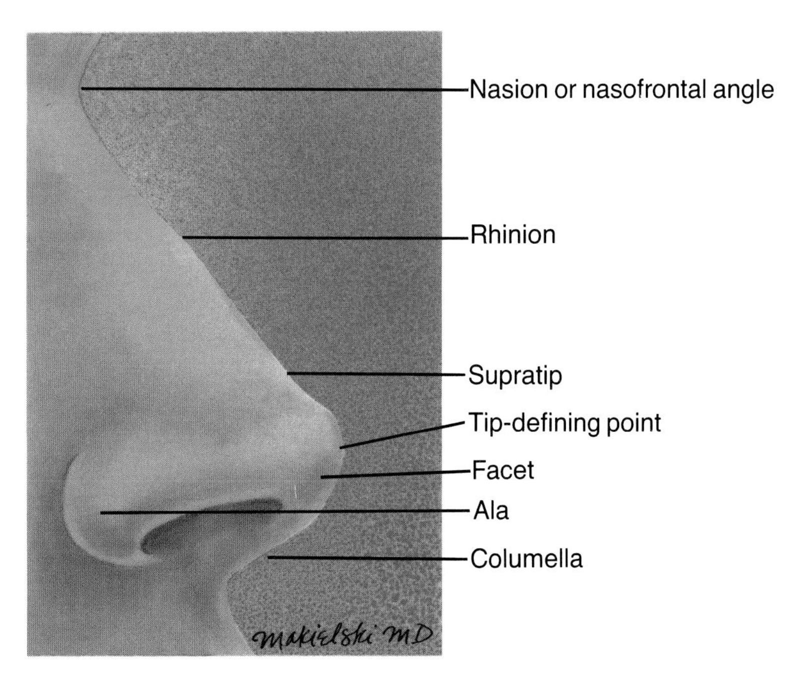

FIG. 14.3. Nasal landmarks: lateral view. A slight fullness at the rhinion and a subtle supratip depression are frequently desirable. In thin-skinned individuals, a delicate external soft-tissue triangle is seen. The ala has a gentle arch superiorly and is neither "hanging" nor retracted. The columella is somewhat visible, has a slight double break, and creates an appropriate angle with the lip.

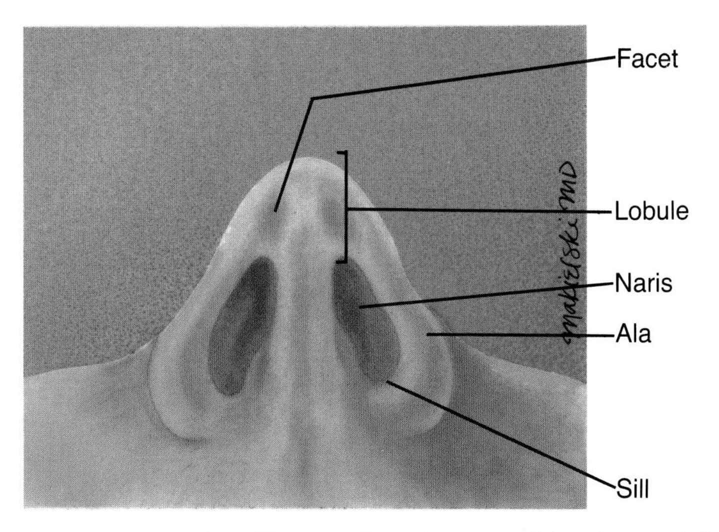

FIG. 14.4. Nasal landmarks: base view. The pear-shaped contour of the alar margin is seen. The relative sizes of the lobule and the ala can be appreciated. The soft-tissue facets are delicate with a smooth margin. The nostril sill creates a gentle curve where the ala fades into the nasal base.

FIG. 14.5. Brows should follow a smooth curve to the nasal dorsum.

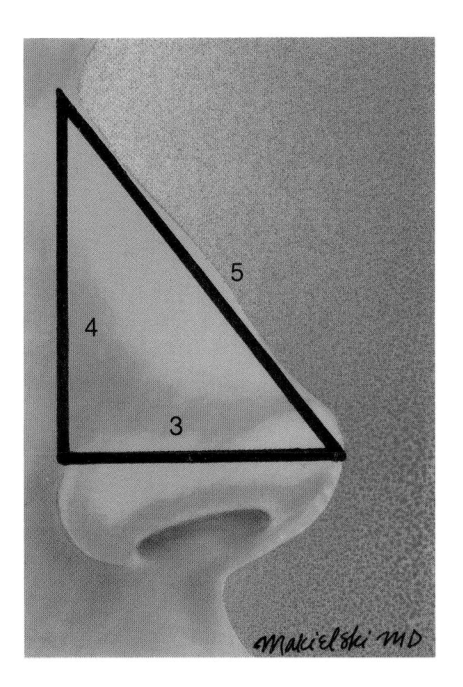

FIG. 14.6. Nasal tip projection: 3-4-5 triangle. A 3-4-5 triangle describes good tip projection. A line from the nasion to the tip is the hypotenuse; a second line is drawn from the nasion to the alar crease and connected with a perpendicular line from the tip. This simple analysis does not consider the relationships of nasal projection to the lip height and the mandible. (Modified from Crumley and Lancer.)

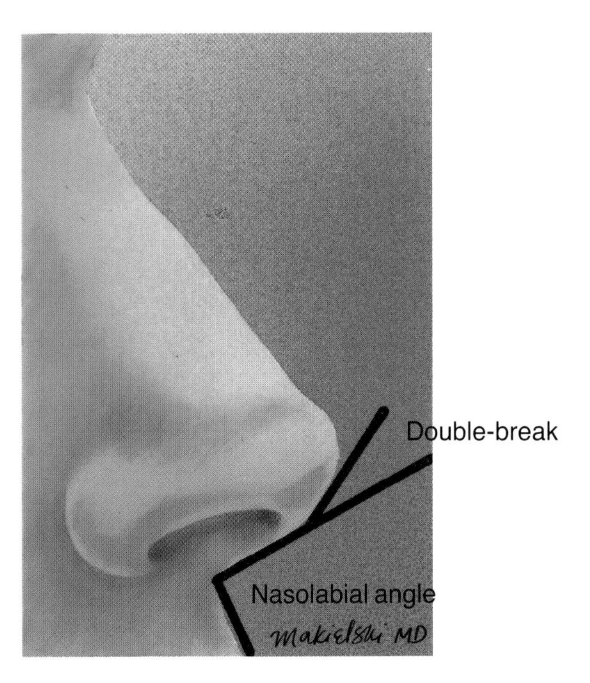

FIG. 14.7. Nasolabial angle and double break. From a lateral view the columella contains a double break, which corresponds to the midseptal angle.

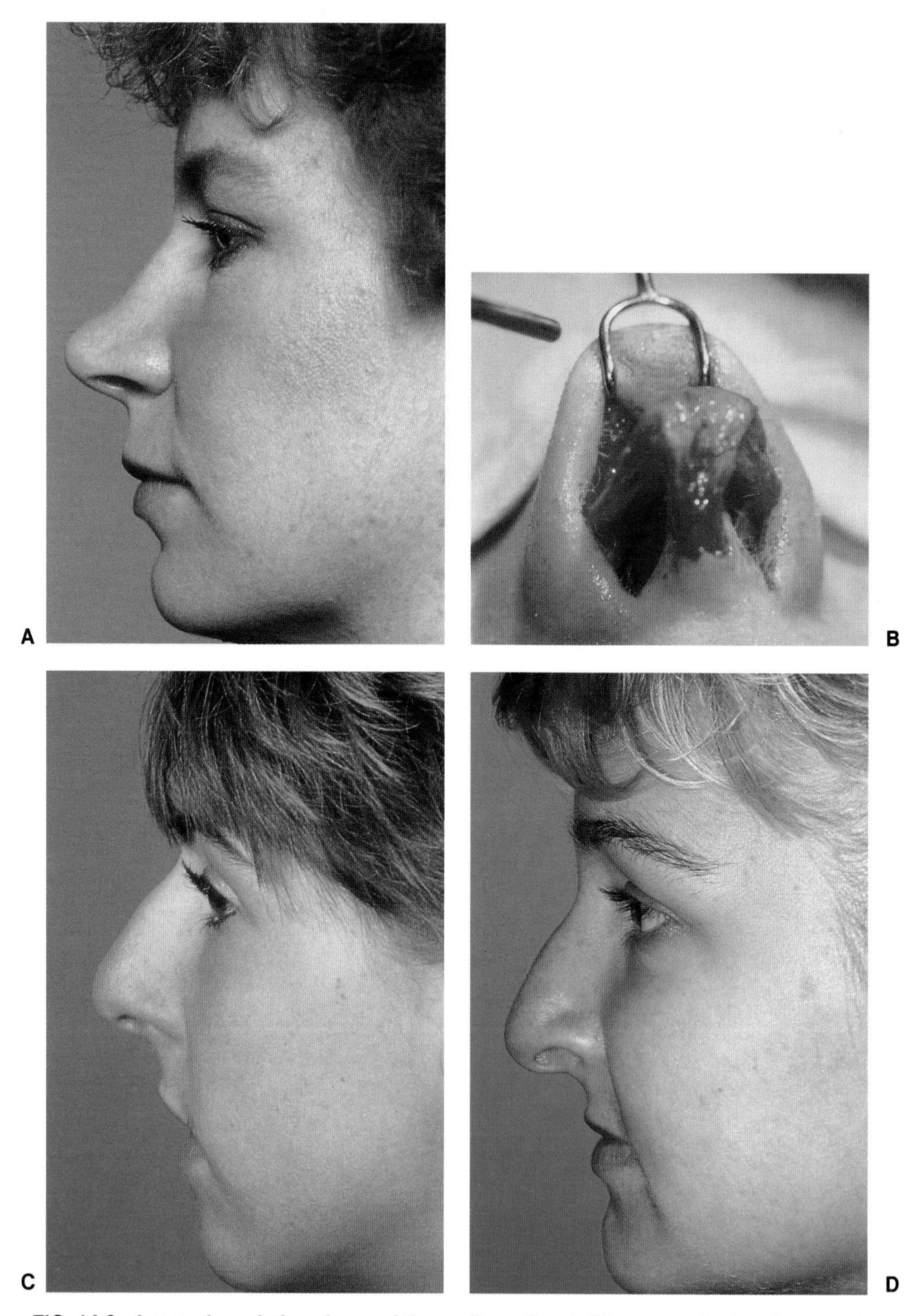

FIG. 14.8. Anatomic variations in nasal tip configuration. A: The overprojecting tip secondary to prominent lower lateral cartilage. **B:** Operative view of the tip cartilages in **A. C:** Blunting of the nasolabial angle secondary to a prominent nasal spine. **D:** The aesthetically important relationship between the columella and the ala is determined by the relative position of the ala (retracted or "hooded") and the columella (retracted or "hanging"). In this patient a relative prominence of the columella is caused by both alar elevation and columellar prominence. The alar position is

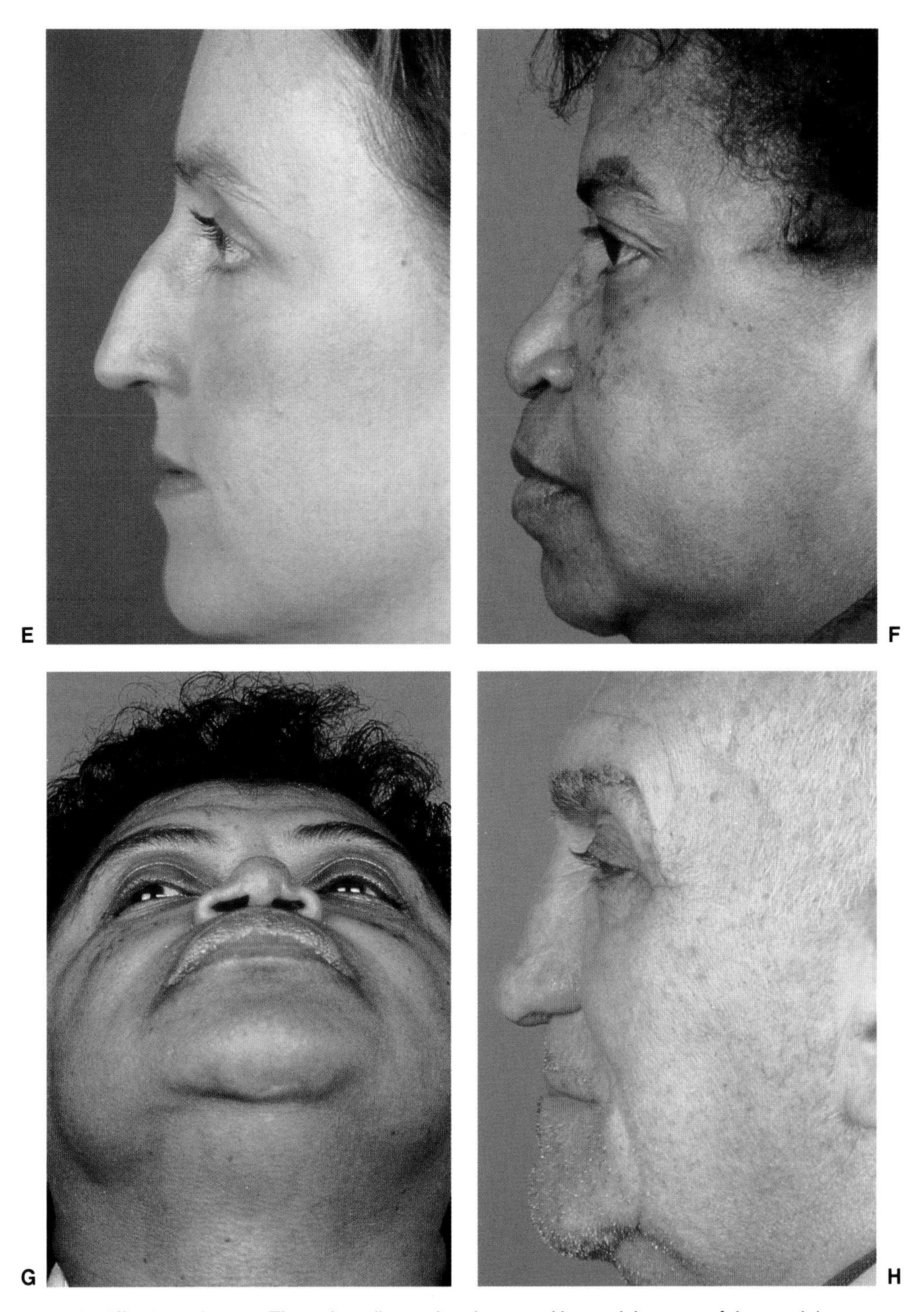

usually difficult to change. The columella can be shortened by excising part of the caudal system and/or medial crura or lengthened with cartilaginous grafts. **E:** A plunging nasal tip secondary to a long lateral crura. **F:** A short columella with an alar-lobular disparity. **G:** Basal view of (**F**) demonstrating a columella-lobular disparity. The lobule in this individual is actually larger than the columella, whereas the columella is normally twice as long. **H:** Loss of tip support mechanisms with age result in a longer nose and a more acute nasolabial angle.

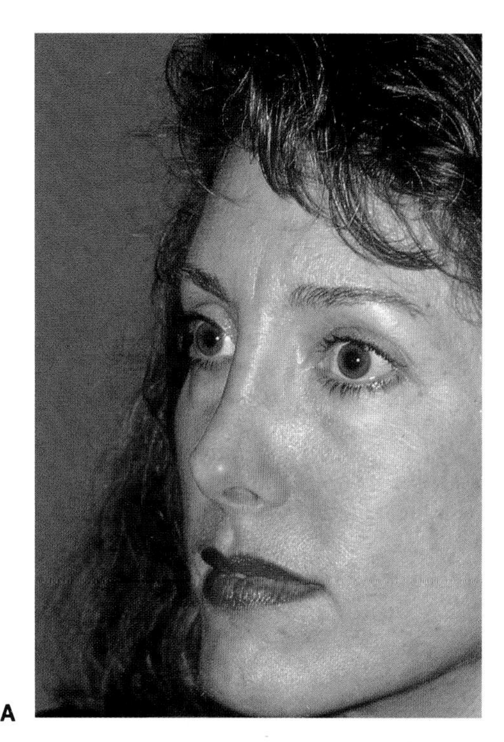

A

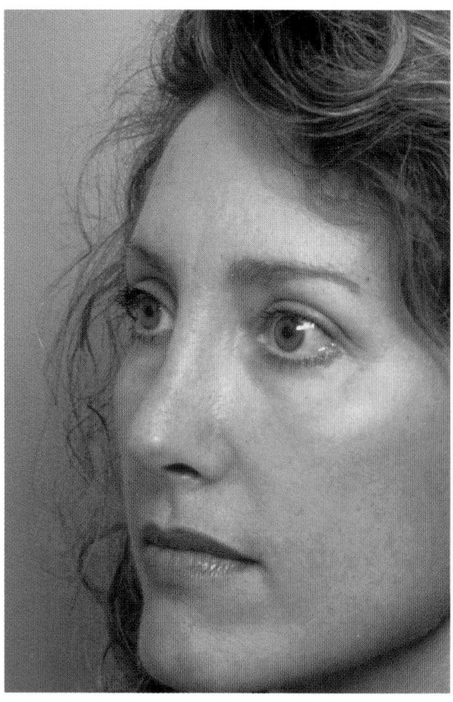

B

FIG. 14.9. Oblique view of the nose. The oblique view is essential for facial analysis and outlines the smooth line from the brow to the nasal tip area. Patient is seen before (**A**) and after (**B**) rhinoplasty and malar implants.

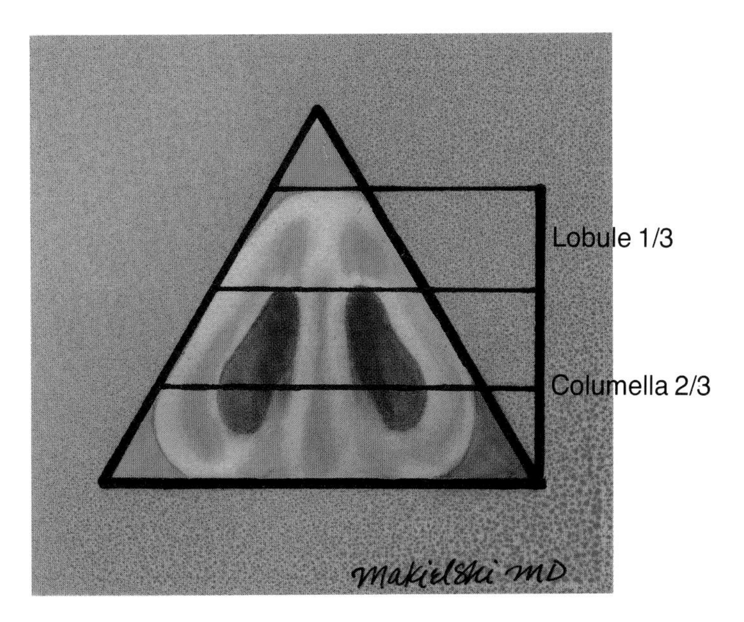

FIG. 14.10. Nasal proportions: base view. The columella is usually twice the length of the lobule.

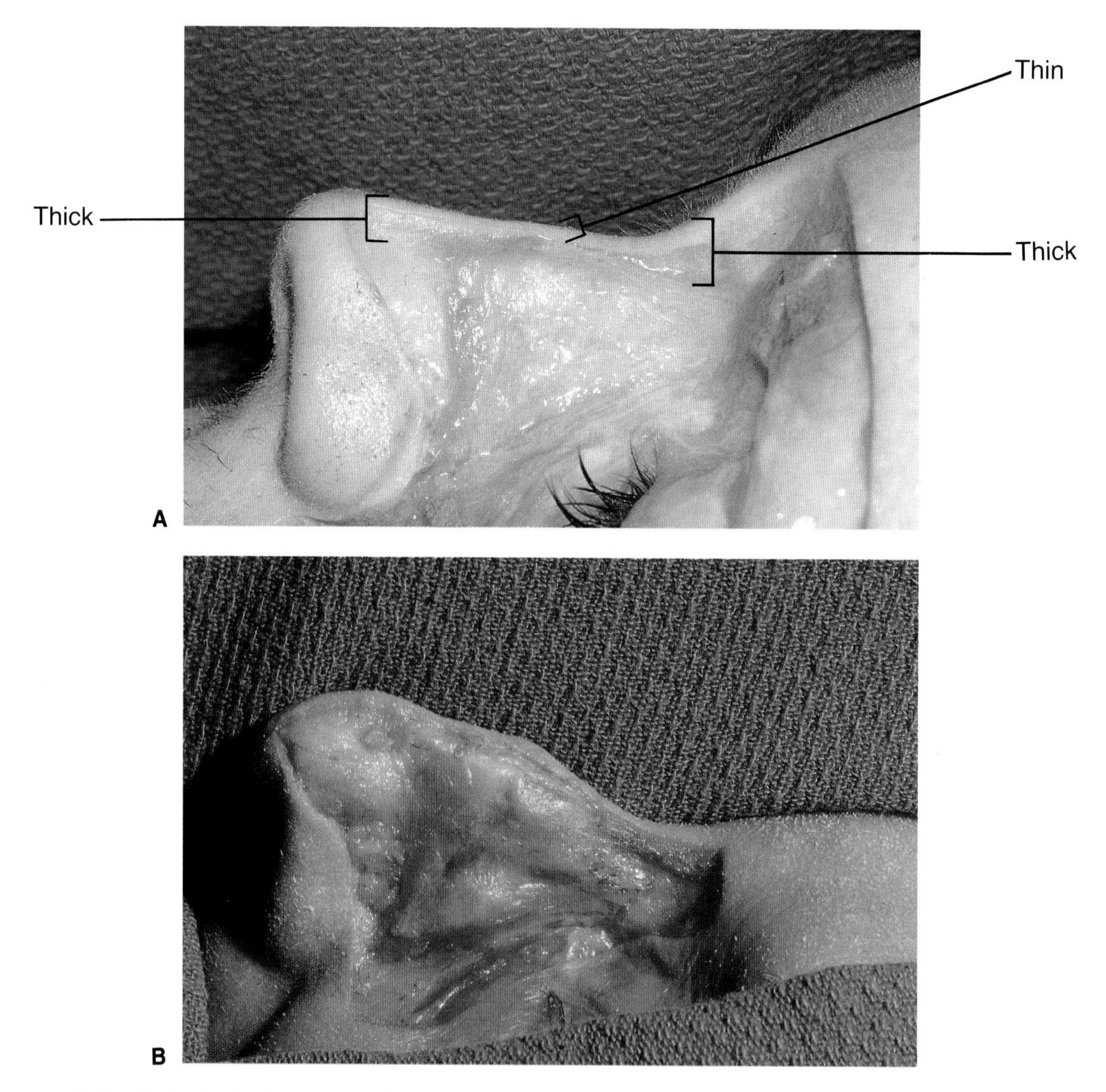

FIG. 14.11. Skin thickness of the nasal dorsum. A and B: The nasal skin thickness varies markedly from patient to patient but is in general thicker in the supratip and the nasion areas. Reduction of the bony/cartilaginous framework in rhinoplasty must allow for variations in the thickness of the skin–soft-tissue envelope. To maintain a straight soft-tissue profile, the framework must maintain a slight hump at the rhinion.

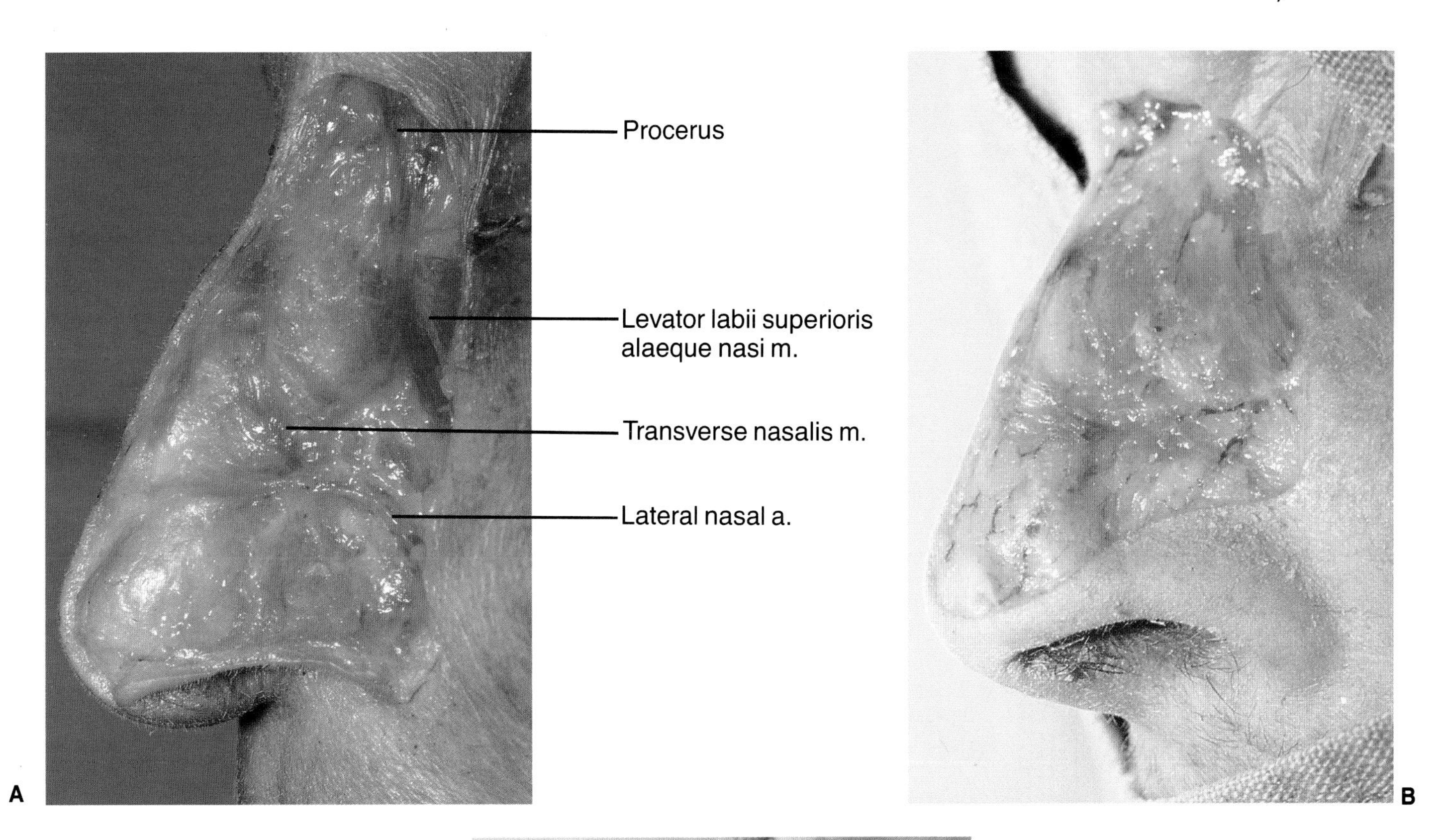

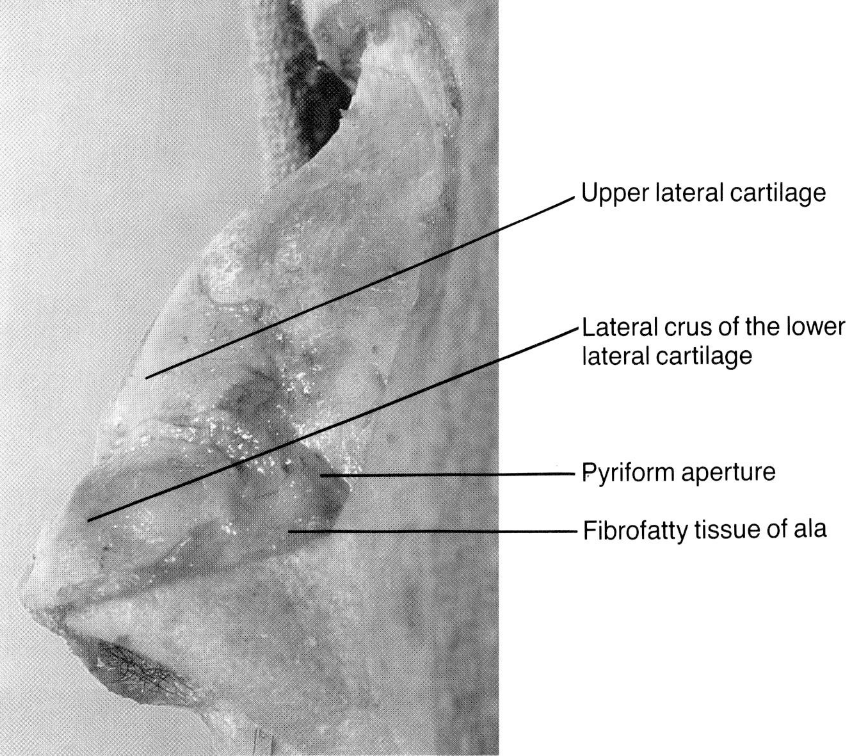

FIG. 14.12. SMAS of the nose. A: The SMAS layer overlying the bony and cartilaginous framework and enmeshing the nasal muscles. **B:** Lateral view of the thinner SMAS in an older individual; the ala is primarily soft tissue. **C:** The SMAS is removed to demonstrate the underlying framework. The upper and the lower lateral cartilage have separated slightly. There is no attachment between the cephalic end of the lateral crus and the pyriform aperture.

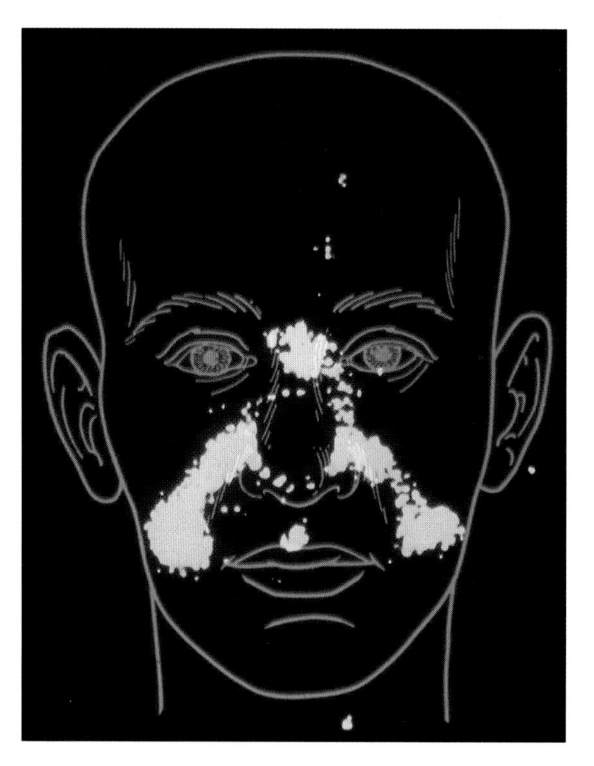

FIG. 14.13. Lymphatic drainage of the nose. Nasal lymphatic drainage follows the venous drainage to the submandibular nodes. Dissection beneath the SMAS and in the midline preserves these drainage patterns. This figure demonstrates the lymphoscintigraphy pattern (yellow) after injection of radionucleotide in the subcutaneous tissue of the nasal tip. There is minimal flow across the columella, and therefore the transcolumellar open rhinoplasty incision does not cause significant postoperative edema. (Courtesy of Dean Toriumi, M.D.)

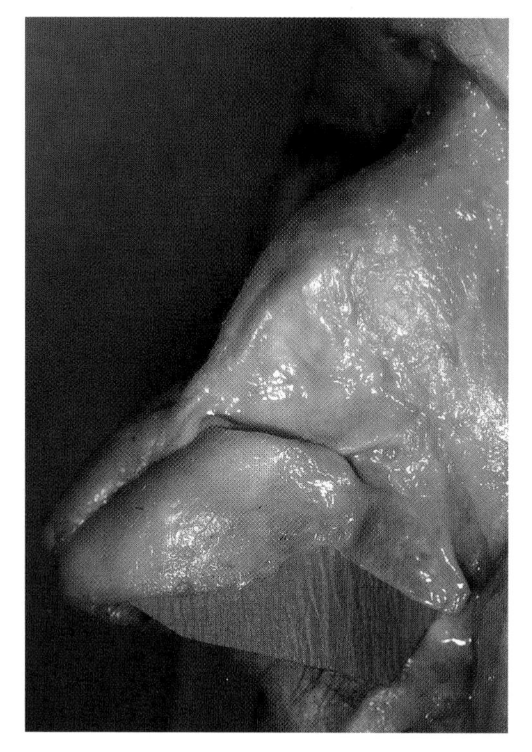

FIG. 14.14. Nasal framework. Note the sesamoid cartilage and the inclination of the lateral crura to the alar margin (blue).

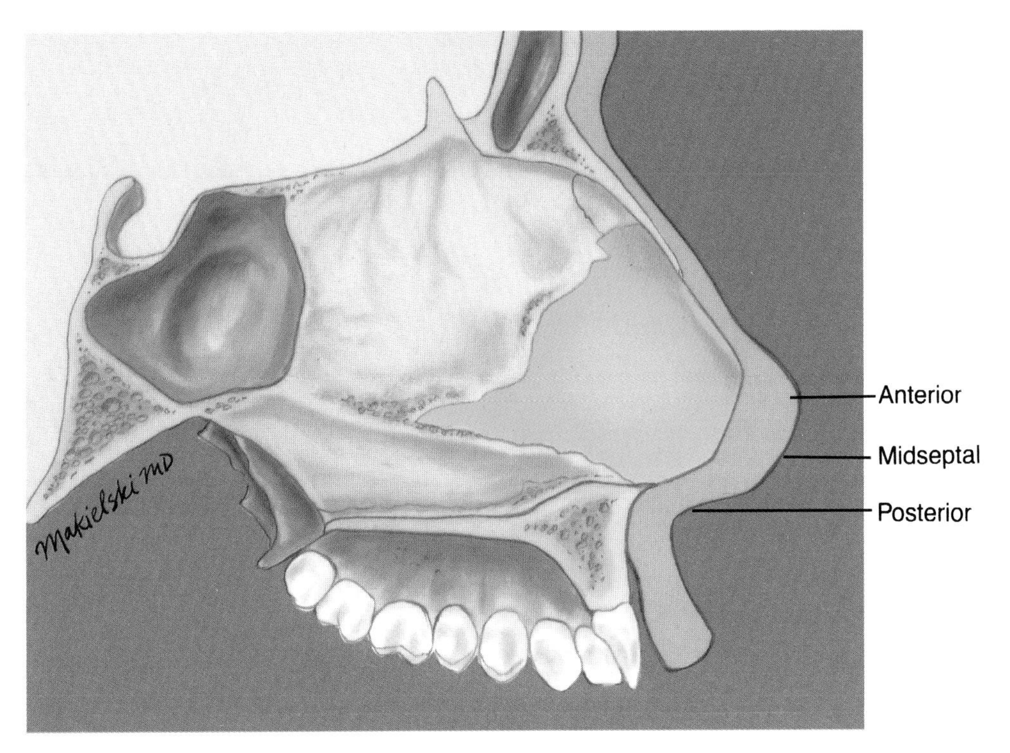

FIG. 14.15. Angles of the nasal septum. The nasal septum has three angles that must be preserved in surgery to maintain the surface contour. These are the anterior, the midseptal, and the posterior septal angles.

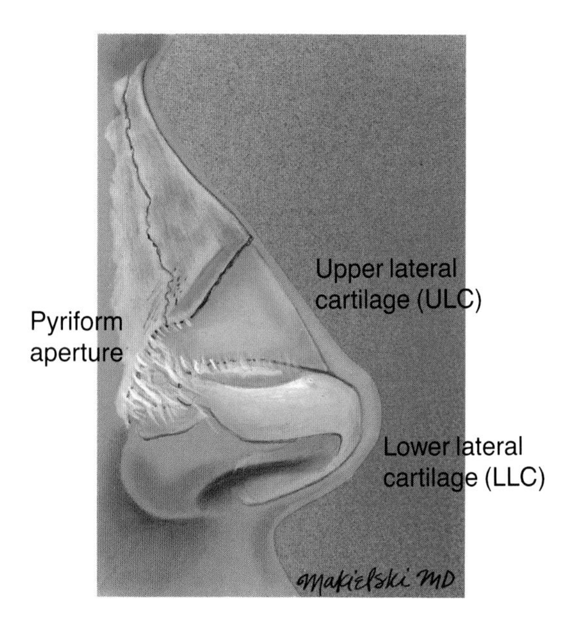

FIG. 14.16. Nasal tip support: lateral view. The intrinsic strengths of the lower lateral cartilage and its attachments to the pyriform aperture, the upper lateral cartilage and the septum provide support to the tip.

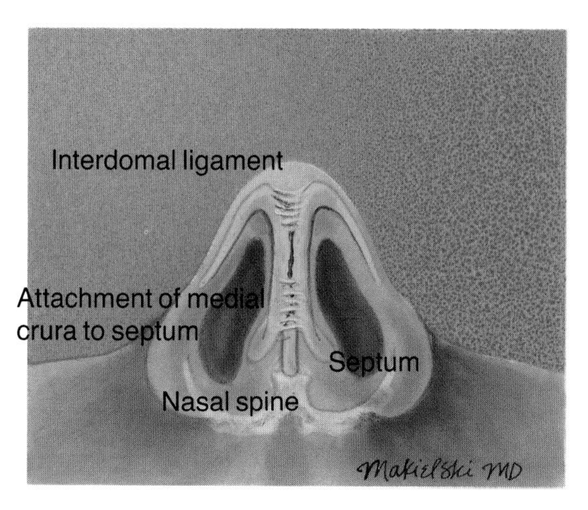

FIG. 14.17. Nasal tip support: base view. From the base view further support mechanisms are seen, including the interdomal ligament, the bony foundation (maxillary crest and nasal spine), and the attachment of the medial crura to the caudal septum.

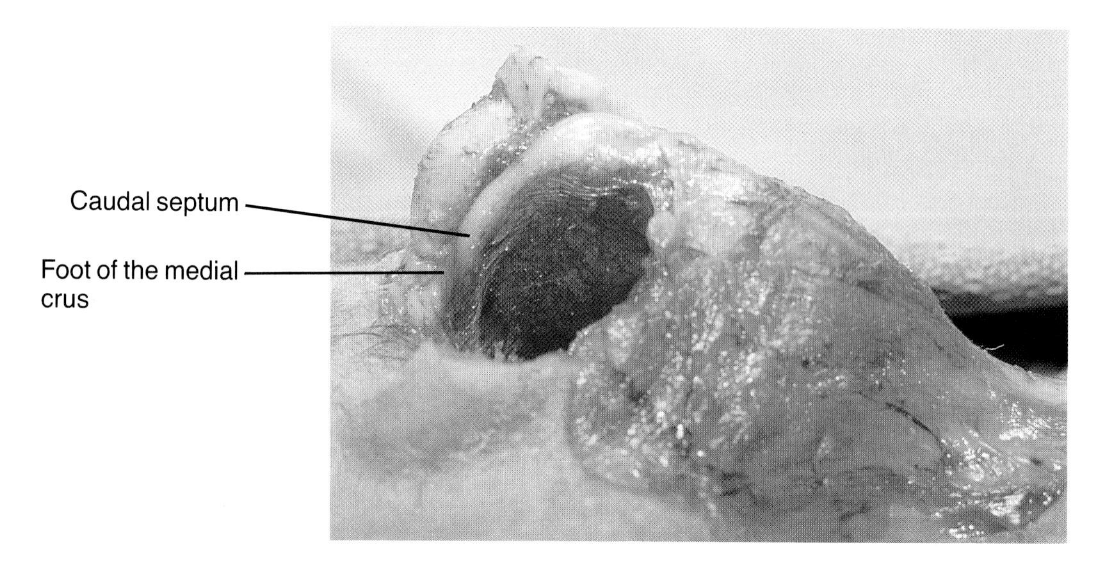

FIG. 14.18. Nasal tip support: crura to septum. The relationship of the medial crura to the caudal septum is one of the tip support mechanisms.

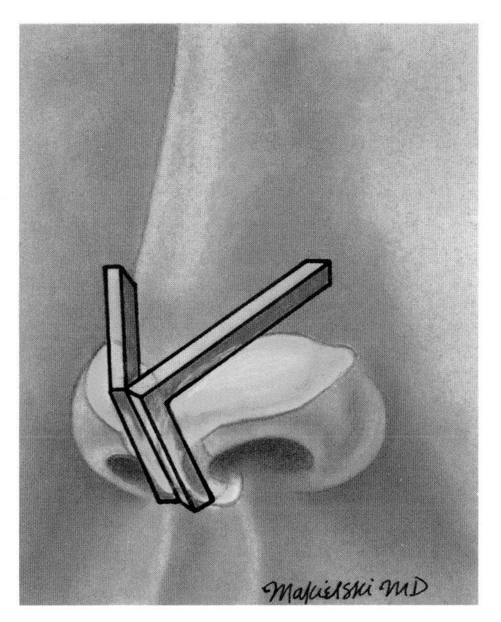

FIG. 14.19. Anderson's tripod concept of nasal tip support. As described by Anderson, the nasal tip can be likened to a tripod with the lateral crura representing two legs and the conjoined medial crura a third. By shortening the two superior legs of the tripod, one can increase rotation and decrease projection. By shortening the conjoined medial crura, one decreases projection and rotation. By cutting across the two lateral crura, one can change the pivot point of the tripod and thus increase rotation without decreasing projection. Selective increase or decrease in the length of the tripod legs can be used to attain the desired tip projection and/or rotation.

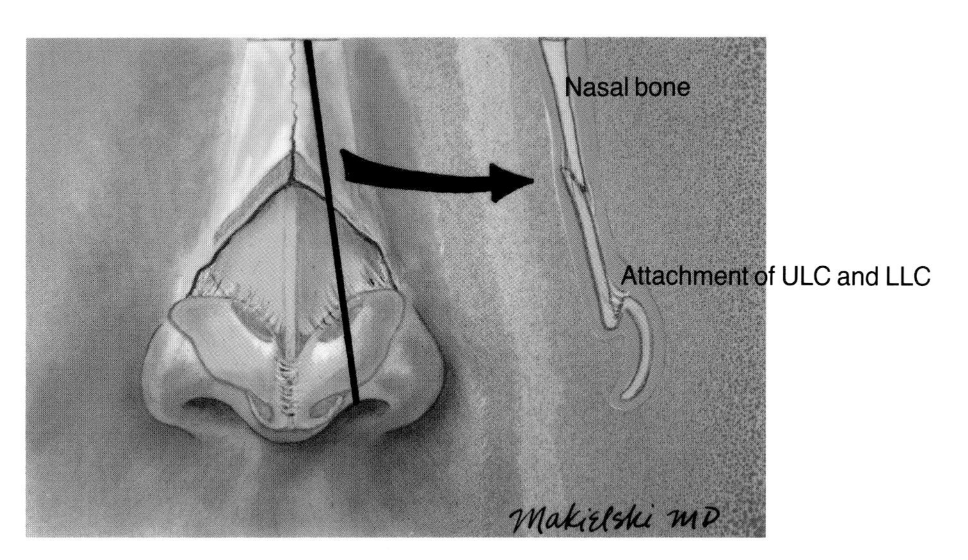

FIG. 14.20. Relationship of the upper lateral cartilage to the nasal bone. Cross-section demonstrating attachment of the upper lateral cartilage (ULC) to the undersurface of the nasal bones and the overlap of the upper and the lower lateral cartilage (LLC). Care must be taken when rasping or otherwise manipulating the nasal bones so as not to detach the upper lateral cartilage from the undersurface of the nasal bones.

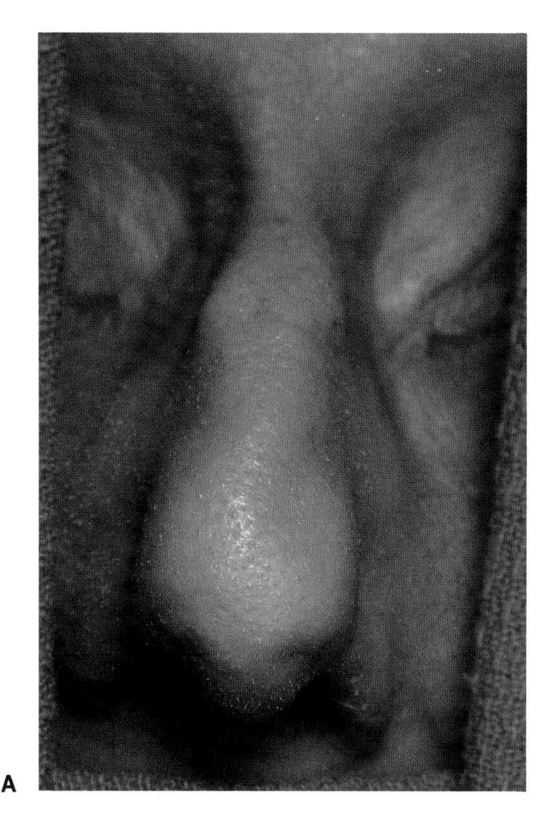

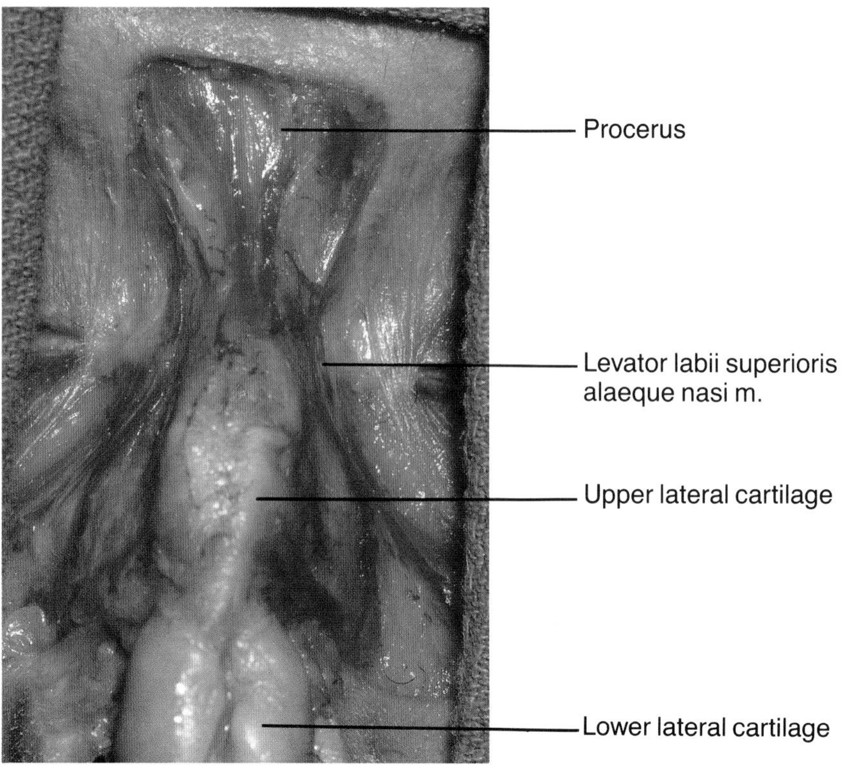

FIG. 14.21. Nasal dissection. A: The external nasal contour is defined by the draping of the skin–soft-tissue envelope over the bony/cartilaginous framework. **B:** The nasal skeleton is exposed, giving an anterior view of the lower lateral cartilage, the upper lateral cartilage, the procerus, and the levator labii superioris alaeque nasi.

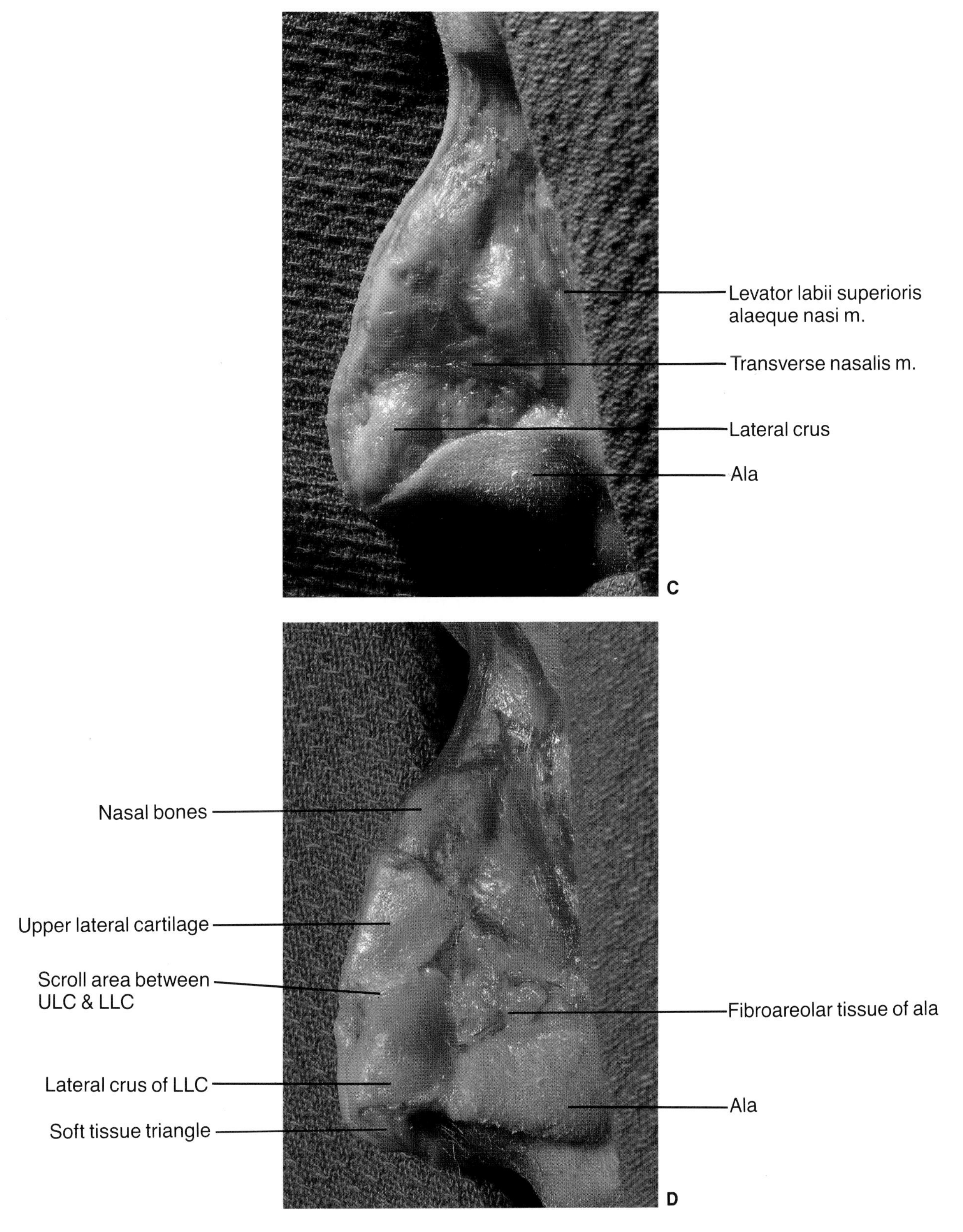

FIG. 14.21. *Continued.* **C:** The ala is primarily soft tissue. **D:** Lateral view of the nasal framework: nasal bones, fibroareolar tissue, ala, soft-tissue triangle, lateral crus, scroll, upper lateral cartilage.

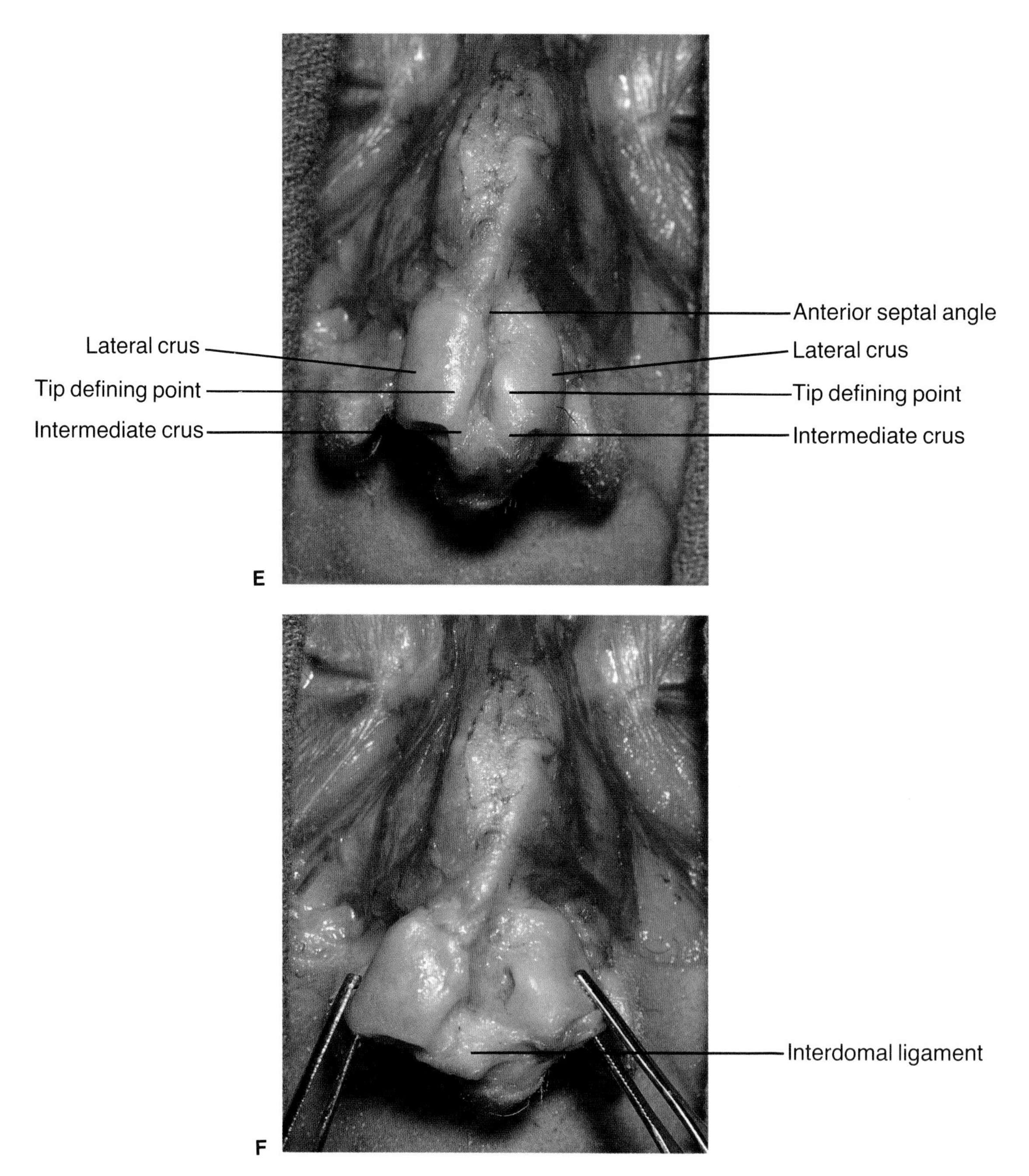

FIG. 14.21. *Continued.* **E:** The curvature of the lower lateral cartilage primarily defines the nasal tip: lateral crus, dome or tip-defining point, intermediate crus, anterior septal angle. **F:** The interdomal ligament helps to support the tip.

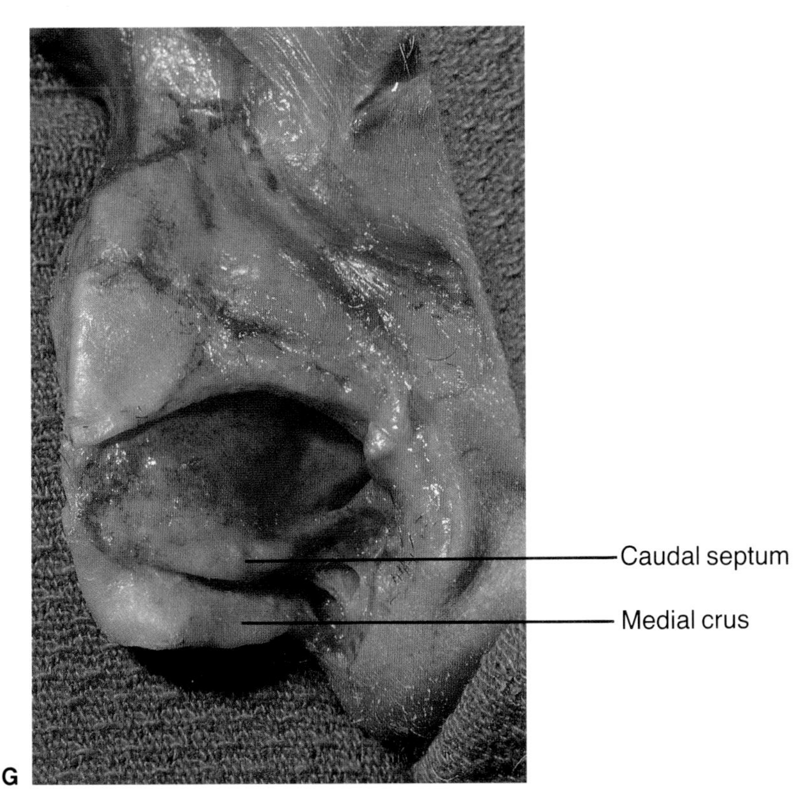

FIG. 14.21. *Continued.* **G:** The left lower lateral cartilage has been removed to allow visualization of the relationship between the right medial crus and the caudal septum.

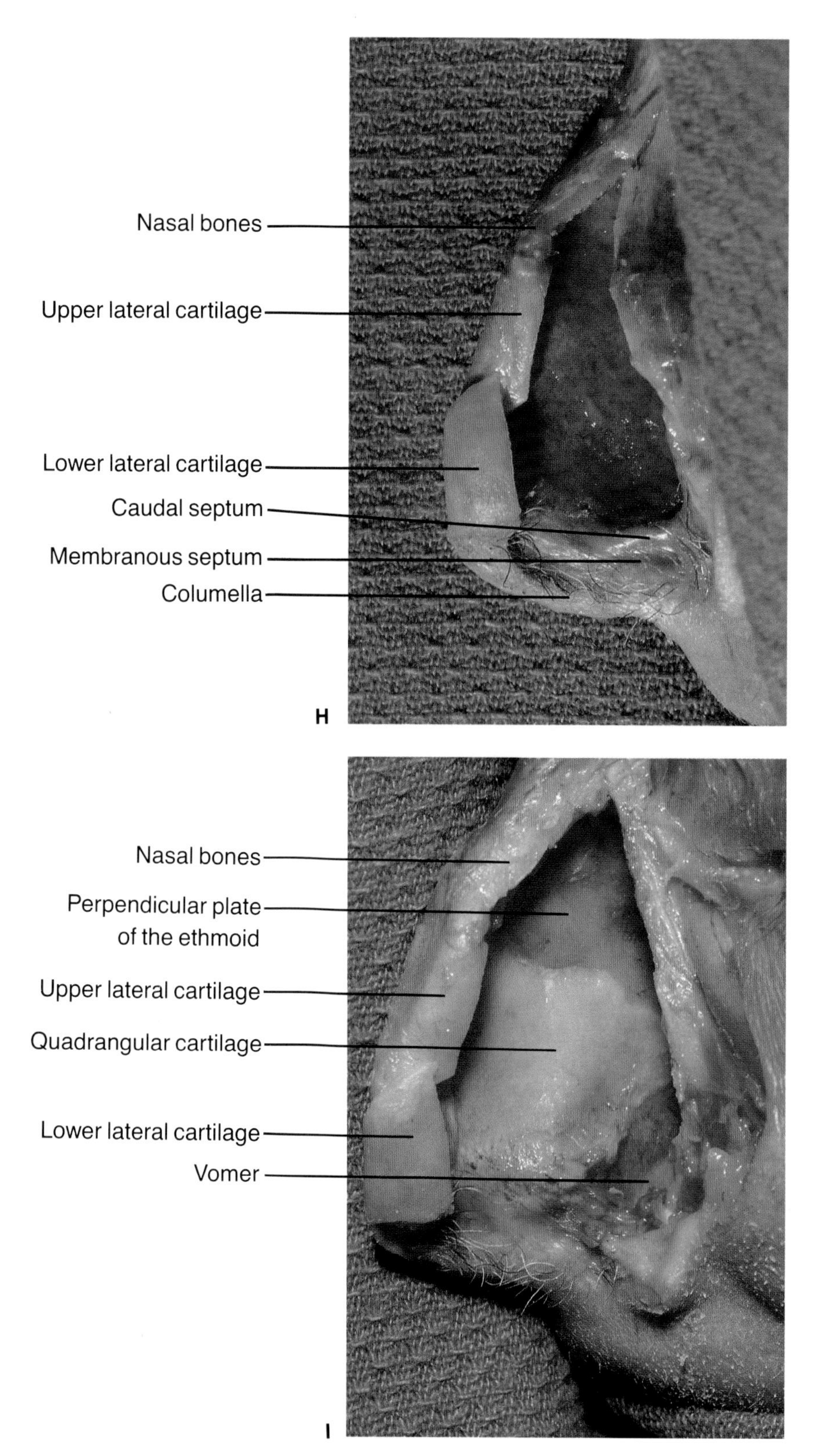

FIG. 14.21. *Continued.* **H:** Relationship of the nasal bones, the upper lateral cartilage, and the lower lateral cartilage to the septum. Also visible are the caudal septum, the membranous septum, and the columella. **I:** Septal mucosa is removed to demonstrate the junction of the bony and cartilaginous septum and the nasal bones, the perpendicular plate of the ethmoid, the quadrangular cartilage, and the vomer.

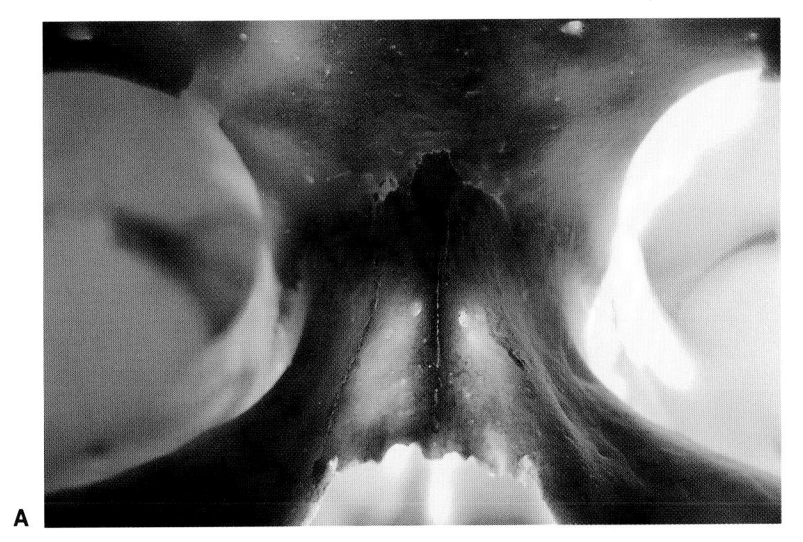

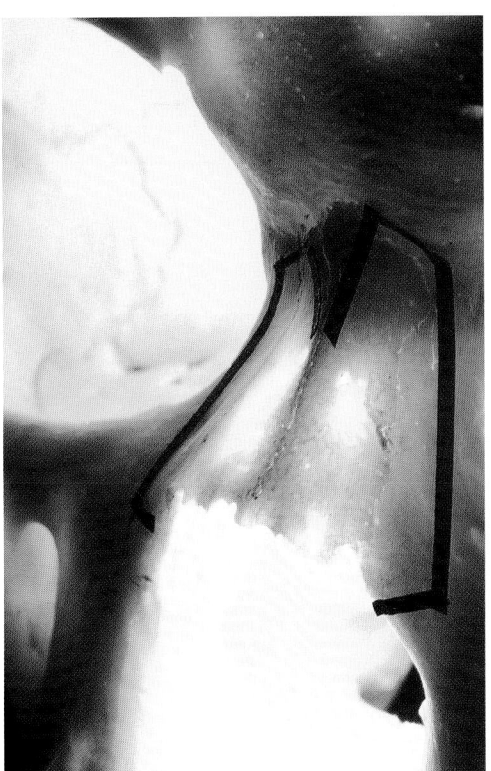

FIG. 14.22. Nasal bones. A: Paired nasal bones and the ascending process of the maxilla create the nasal bony pyramid. A transluminated skull shows the thicker nasal bone cephalically and the thinner bone caudally. **B:** Incorrect osteotomies (left) carried high into the thicker bone can create a "rocker" in which a transverse fracture occurs in the frontal bone; pressure on this superior fracture site will then cause the lower nasal bone to protrude. This problem is remedied by performing a percutaneous osteotomy with a 2 mm osteotome at the junction of the thick and the thin nasal bones.

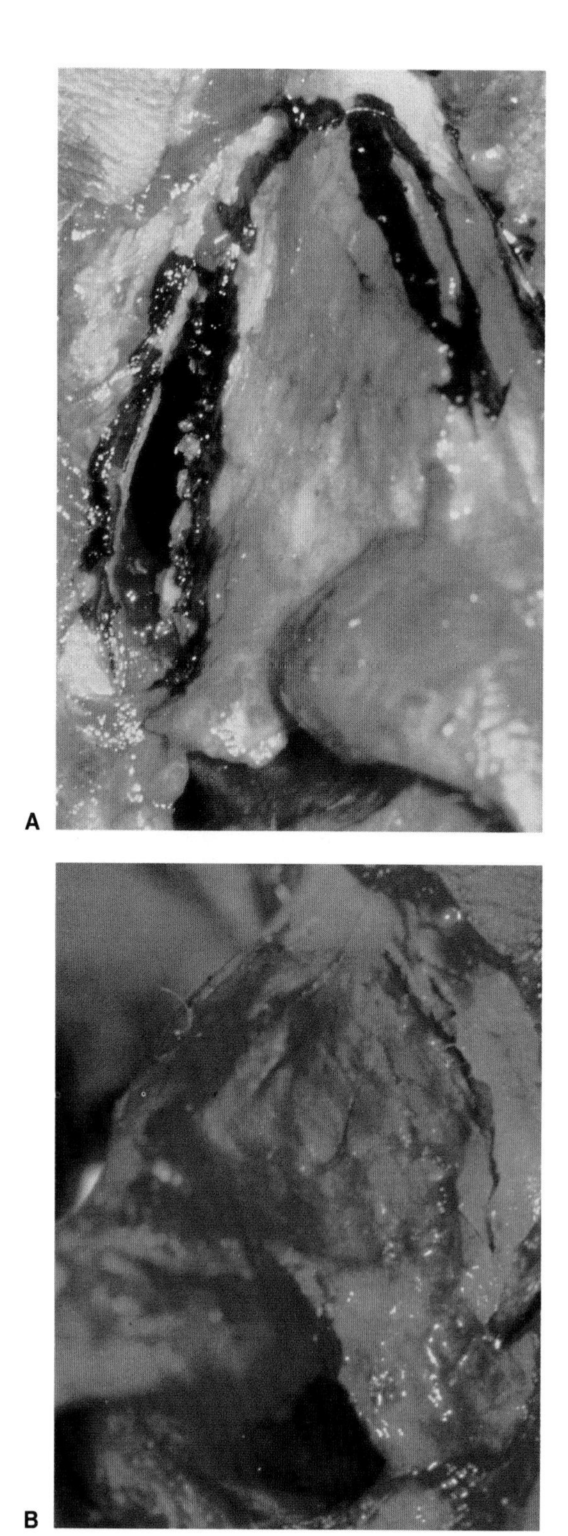

FIG. 14.23. Lateral osteotomies of the nasal bones. A: An osteotomy created with a guarded Neivert osteotome shows a smooth cut, but some loss of soft-tissue support. **B:** An osteotomy created with microperforations using a 2 mm osteotome shows a slightly irregular cut with good soft-tissue support.

CHAPTER 15

Ears

The external ear serves as a sound collection device and creates an acoustic shadow that is extremely important in sound orientation and localization. The auricle also provides external protection for the more delicate middle and inner ear structures.

AESTHETIC PROPORTIONS

The major landmarks of the external ear are presented in Figure 15.1. The usual proportions for adult Caucasians are approximately 6.5 × 3.5 cm (Fig. 15.2); African ears are generally shorter, whereas Asian ears are generally longer. There is great individual variation in appearance, but the following guidelines are generally true of aesthetically pleasing ears.

The external ears should appear bilaterally symmetric when viewed from the front, with emphasis on perfect symmetry along the lateral helical rim. Slight asymmetries of the lobule and the antihelix are less noticeable than those involving the helix. In the frontal view, the helical rim should be visible throughout its entire length. The antihelical fold should angle forward superiorly. The antihelix should form a smoothly curved ridge, curving more strongly inferiorly where it blends into the antitragus.

Superiorly, the inferior crus curves anteriorly and forms the only region where the antihelix appears "sharp." The superior crus fans out gently into the region of the fossa triangularis, which should face out in the lateral direction. The antihelix should be at nearly a right angle to the concha cartilage, and the floor of the concha should parallel the mastoid surface.

The angle between the superior aspect of the helix and the mastoid plane is most aesthetic at 20° to 30°, which usually translates into a measurement of 15 to 20 mm from the helix to the mastoid surface (Fig. 15.3). Abnormal protrusion occurs in approximately 5% of the Caucasian population and correlates with angulation in excess of 30° to 40°. Relative protrusion of the middle portion of the ear is more acceptable than protrusion of the superior pole and the lobule. There should be no excess tubercle of Darwin.

EMBRYOLOGY AND DEVELOPMENT

Embryologically, the auricle develops along the first branchial groove from the first and second branchial arches. The external ear is formed from six auricular hillocks in the neck and later migrates to the side of the head. Congenital abnormalities, e.g., inferior displacement of the ear and preauricular cysts, are related to this embryological development.

The ear continues to grow following birth and attains approximately 85% of its adult size by 3 years of age. Slow growth may continue until approximately 6 years of age, at which time the ear has reached an average adult size of approximately 6.5 × 3.5 cm (Fig. 15.2). The distance from the mastoid periosteum to the helical rim changes little after 10 years of age, although the ear may continue to elongate more than 1.5 cm along its vertical axis during the lifetime of the individual.

ANATOMY

Cartilage

The shape of the auricle stems from a single piece of elastic cartilage (Fig. 15.4). The cartilaginous framework follows the external contour of the ear with the notable exception of the lobule, which lacks any cartilaginous framework (Fig. 15.5). The cartilage forms a nearly complete circle around the auditory meatus; the gap between the tragus and the crus of the helix is bridged by a ligament. Fissures often exist in this gap, allowing for bidirectional transmission of tumor or infection between the mastoid or the parotid regions and the external canal. Thin skin adheres tightly to the lateral aspect of the auricle; the skin on the medial aspect is thicker and looser.

The medial aspect of the conchal cartilage approximates the mastoid bone and acts as the main buttress of the ear, holding the auricle away from the head. The antihelix is a smooth, ridge-like elevation of the cartilage encircling the lateral conchal margin. Superiorly, the antihelix divides into a superior and an inferior crus and crosses a shallow depression known as the fossa triangularis. Between the antihelix and the surrounding helical rim lies the crescent-shaped region of the scaphoid fossa. The anterior helical crus partially divides the conchal cavity into the cymba conchae and the cavum conchae. Inferiorly, the antihelix is continuous with the lateral external auditory cartilage by way of a narrow cartilaginous band, the isthmus, which in turn also bears the tragus.

Muscles and Ligaments

Each ear has six intrinsic and three extrinsic muscles that lend minor structural support to the auricle (Fig. 6.6). These muscles are innervated by branches of the seventh cranial nerve.

The basic position of the external ear results from the inherent elastic properties of the ear cartilage, with added support coming from a number of ligaments. The anterior extrinsic ligament extends from the root of the zygoma to the tragal cartilage and to the spine of the helical crus. The posterior extrinsic ligament attaches the posterior aspect of the concha cartilage to the underlying mastoid periosteum.

CLINICAL APPLICATIONS

The Protruding Ear

There are an enormous number of variations in the position, the size, and the shape of the auricle. At least 40 descriptive names in the clinical literature have been applied to these variations. The two most common anatomic components of the protruding ear are the lack of an antihelix (Fig. 15.3) and a prominent concha. An additional common associated deformity is a protruding lobule.

Surgical approaches to these deformities create a new antihelix with either a suture technique or cartilage manipulation and reduce the size of the concha as needed (Larrabee et al., 1992).

The Auricular Cartilage Graft

Autologous cartilage grafting is useful in a variety of clinical settings such as cosmetic rhinoplasty and in reconstructive surgery of the nose, the eyelid, and the auricle. The external ear may serve as a source of graft material in those circumstances where more functional sources of cartilage such as the nasal septum are inadequate or unavailable. The harvesting of graft material is generally limited to the region of the concha. Resection of cartilage should be limited to the flatter medial region of the conchal bowl, avoiding removal of the vertical portion that contributes to the structure of the antihelix. Furthermore, the radix helicus must be left undisturbed in order to assure a pleasing aesthetic shape and to maintain support.

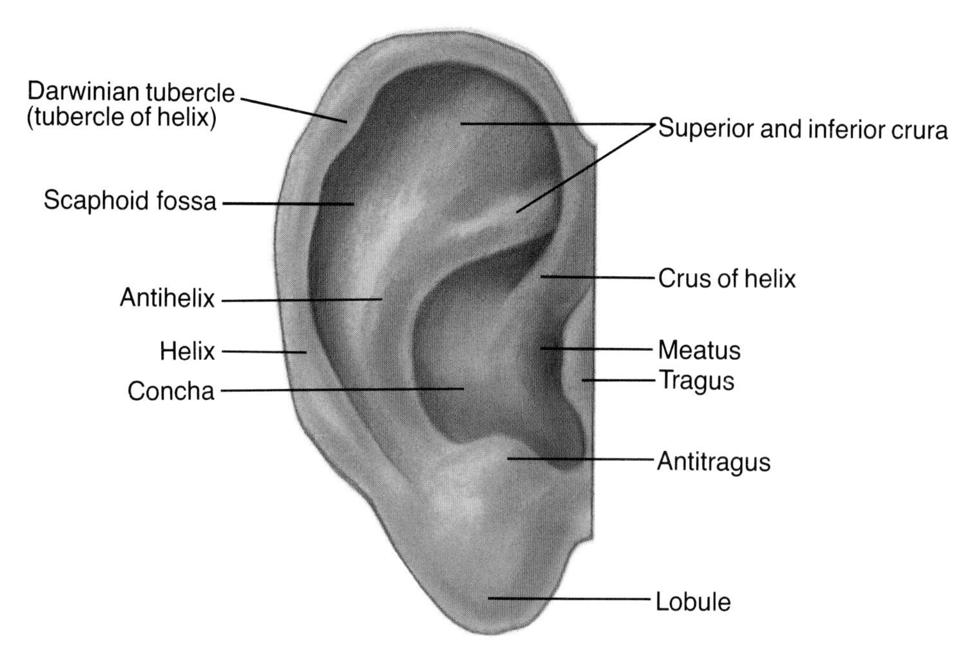

FIG. 15.1. Auricular landmarks. The thin auricular skin conforms closely to the curvature of the underlying cartilage that creates the complex convolutions of the auricle. (The skin is more adherent laterally than medially and therefore composite skin-cartilage grafts are best harvested from the lateral surface.) The helix, antihelix, tragus, and lobule are the major aesthetic landmarks of the auricle. Symmetry of the helix is particularly important from the frontal view. The helical rim should be visible throughout its entire course. It should create a smooth curve and fade evenly into the lobule. The lobule should be neither protruding nor retracted. The antihelix should be a smoothly curved ridge that blends into the antitragus inferiorly. Superiorly, the antihelical fold turns forward into the sharp inferior crus and the flatter superior crus that surround the fossa triangularis.

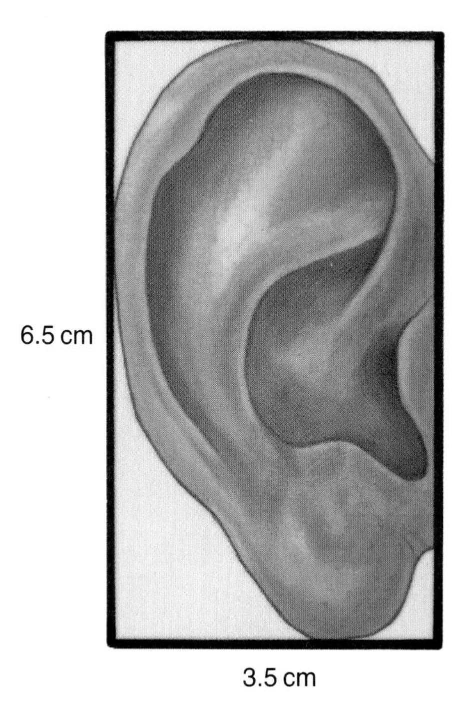

FIG. 15.2. Auricular proportions. The above dimensions are typical, but other measurements are equally important. The long axis of the ear slants posteriorly, creating an angle with the vertical of about 20°. In its vertical length, the ear extends from the brow to the base of the nose and lies about one of these ear lengths posterior to the lateral orbital rim.

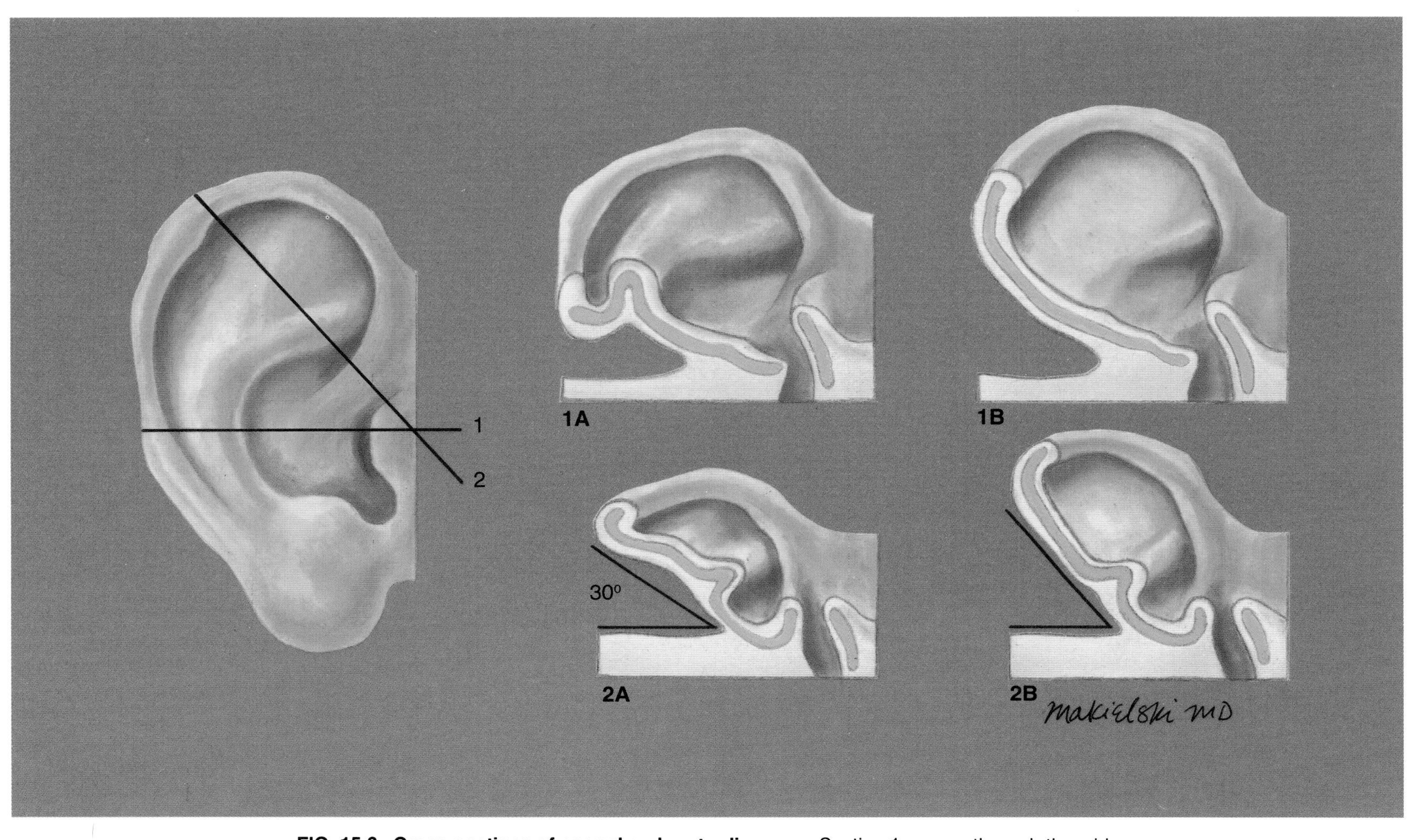

FIG. 15.3. Cross-sections of normal and protruding ears. Section 1 passes through the mid-auricle. **1A:** Shows the normal cartilage configuration. **1B:** Shows a protruding auricle, where the antihelical fold is absent. Section 2 passes through the upper auricle. **2A:** Shows the normal 30° angle between the mastoid and the auricle. **2B:** Demonstrates the wider angle of a protruding auricle.

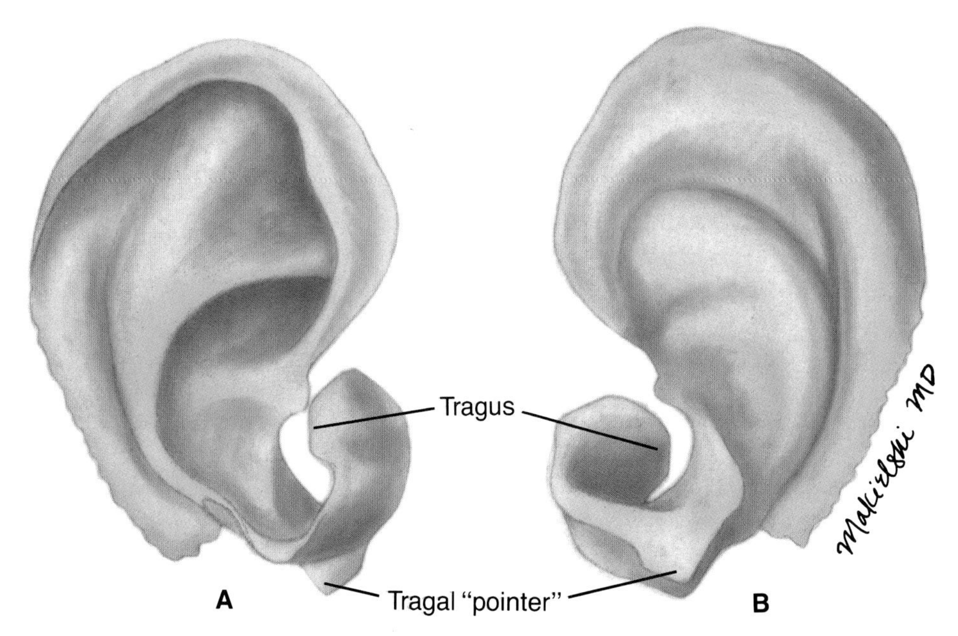

FIG. 15.4. Auricular cartilage. The cartilaginous framework creates the auricle's contour except where it is lacking in the lobule. The tragal (or cartilaginous) "pointer" is a helpful landmark when identifying the main trunk of the facial nerve at the stylomastoid foramen. **A:** Lateral view. **B:** Medial view.

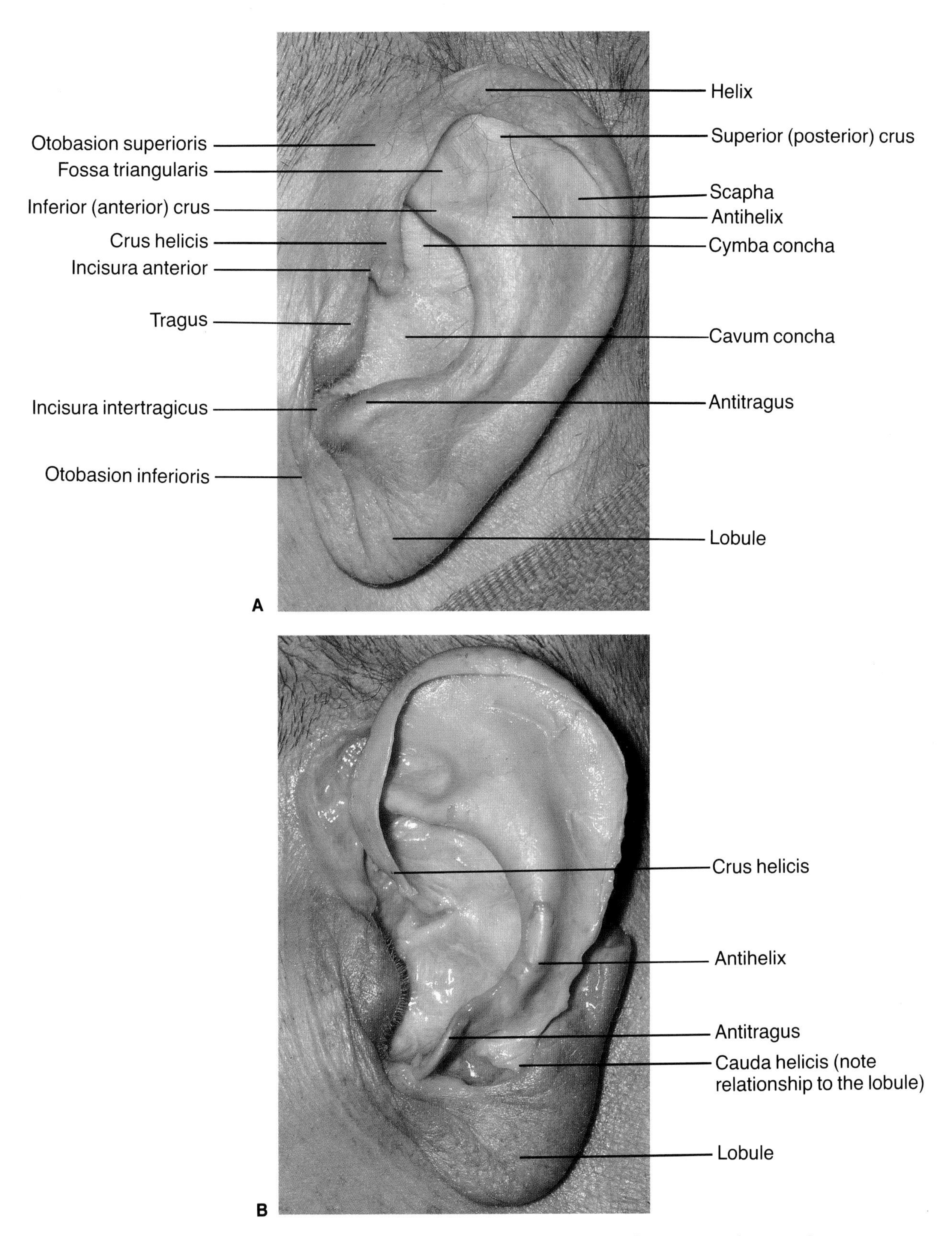

FIG. 15.5. Dissection of the auricle. A: Auricle with intact skin. **B:** Cartilaginous framework with soft-tissue lobule. Note the relationship of the cauda helicis to the lobule; repositioning of the cauda helicis can affect the lobule position in otoplasty.

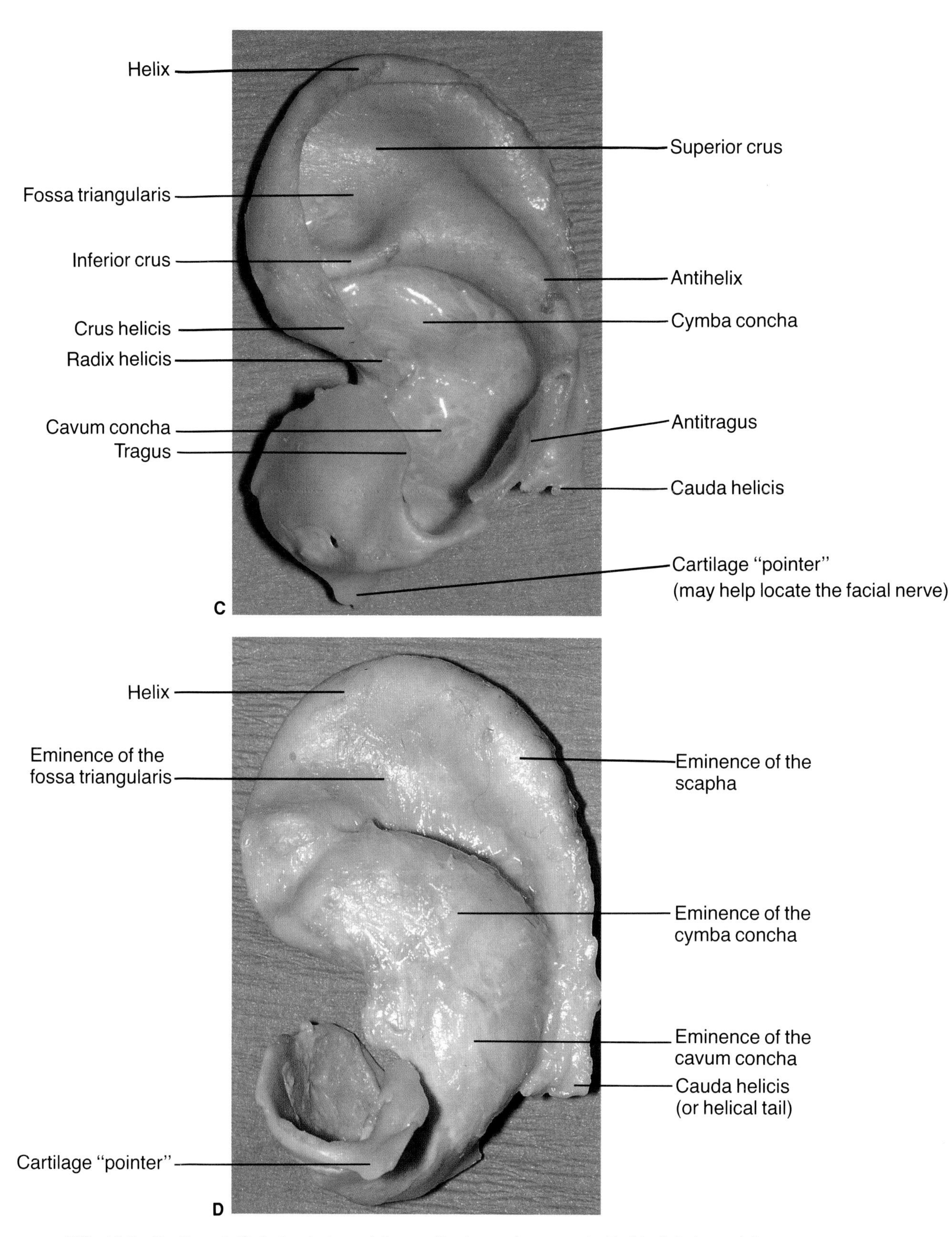

FIG. 15.5. *Continued.* **C:** Lateral view of the cartilaginous framework. **D:** Medial view of the cartilaginous framework.

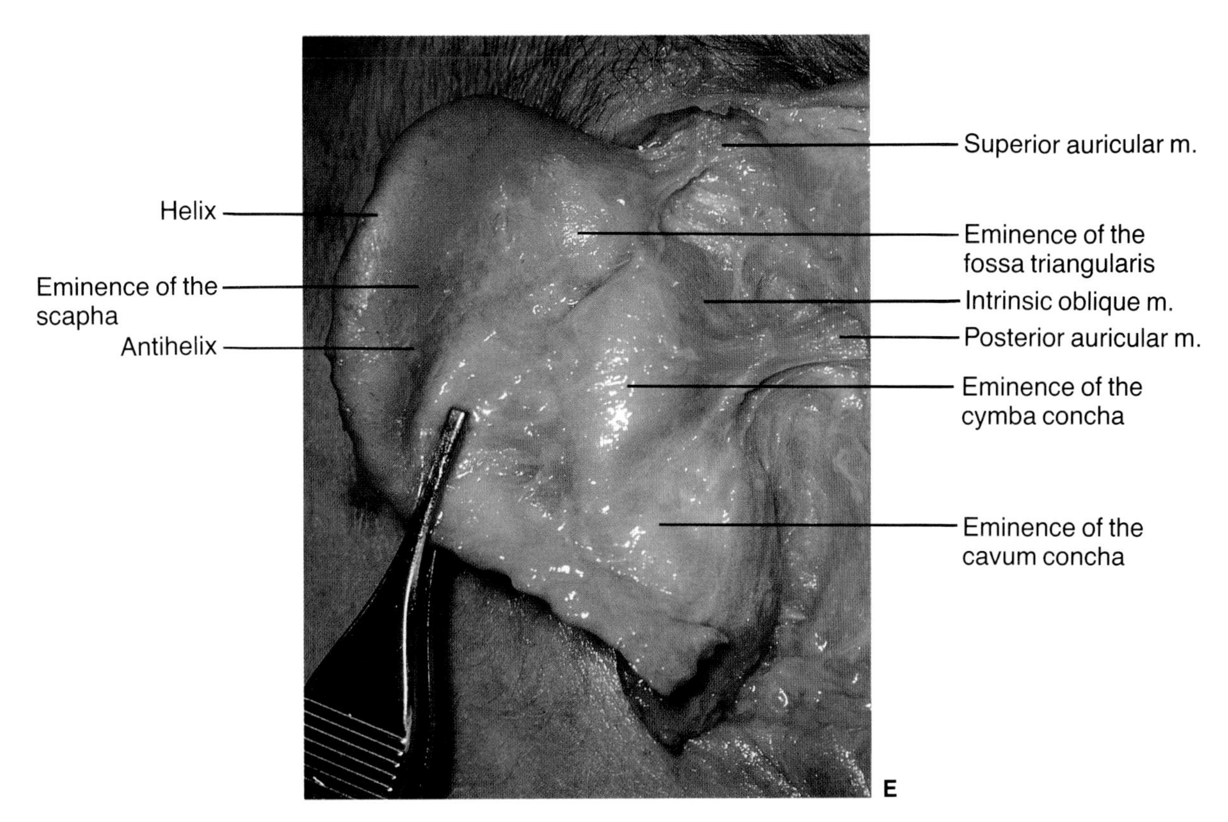

FIG. 15.5. *Continued.* **E:** Posterior view of the auricular musculature.

CHAPTER 16

Cheeks and Neck

The contour of the cheek, unlike the other asethetic units of the face, is defined primarily by soft tissue. This soft tissue is framed by the malar complex above and the mandible below, but its shape is largely determined by the parotid gland, facial musculature, and buccal fat.

The contour of the neck represents a summation of the effects of the bony/cartilaginous framework and the soft-tissue envelope. From the lateral and the oblique views the strength of the mandible defines the shape of the neck. The relative prominence of the sternocleidomastoid muscle and the mandible create an aesthetically important lateral triangle. The cervical mental angle and the profile of the neck are determined by the positions of the mandible, the hyoid bone, and the thyroid cartilage plus the overlying fat, muscle, and skin. The amount and the location of fat, the laxity of the skin, and the anatomy of the platysma muscle contribute to the overall aesthetics of the neck in both young and old individuals.

The most common aesthetic procedure performed on the cheeks and the neck is the cervicofacial rhytidoplasty or face-lift. Face-lift procedures usually involve (1) various degrees of undermining the skin and the subcutaneous tissue, (2) tightening of the underlying SMAS and the platysma (Fig. 16.1), and (3) redraping and excision of excess skin. The majority of fat removed during liposuction of the neck is superficial to the platysma; some surgeons remove fat between its anterior borders.

MALAR COMPLEX

The malar complex is a key component of facial form, but its curves are not easily analyzed. Two methods for determining the aesthetic position for the most prominent part of the malar eminence have been developed. The Hinderer analysis is shown in Figure 16.2; Powell's system is presented in Figure 16.3.

THE BUCCAL FAT PAD

The aesthetically important part of the buccal fat pad lies lateral to the buccinator and provides fullness to the cheek inferior to the malar prominence. The fat pad as a whole is a complex three-dimensional structure. It is a unique form of fat, termed a syssarcosis, whose main function is to provide a surface for the gliding motion of the muscles of mastication. Its anatomy can be best understood in this context.

The buccal fat pad was first accurately described by Bichat in 1801, although Heister

(1732) had previously erroneously reported it as a buccal salivary gland. It lies within the masticatory space. Its weight is remarkably constant at approximately 8 g and does not vary with the overall degree of adiposity of the subject. There is no variation between the sexes or from one side to the other in an individual. It is quite distinct in appearance from facial fat and resembles orbital fat in texture and color. There are few fibrous septae to be found, and on dissection the fat can be teased out with ease in large globules.

The main body of the buccal fat pad sits on the posterolateral surface of the maxilla overlying the superior portion of the buccinator muscle under the anterior portion of the masseter (Figs. 16.4 and 16.5). It has three main extensions: buccal, pterygoid, and temporal, although some authors have divided it into as many as nine portions.

The buccal extension is the most important from an aesthetic standpoint. It extends anteroinferiorly from the main body in a globular mass with a convex lateral face, covering most of the buccinator. Two surgically vulnerable structures, the facial nerve and the parotid duct, lie in close proximity to the buccal extension. The buccal rami of the facial nerve that exit the anterior border of the parotid gland lie on the surface of the masseter, tightly bound by the thin sheet of parotid-masseteric fascia. An extension of this fascia envelops the buccal fat pad; thus, the buccal extension has facial nerve fibers immediately lateral to it. The parotid duct runs immediately superior to the buccal extension before diving through the fibers of the buccinator to enter the oral cavity.

The pterygoid extension continues posteriorly from the main body on the lateral surface of the medial and the lateral pterygoid muscles and proceeds medially around the posterior wall of the maxilla and through the pterygomaxillary fissure. It can be encountered closely associated with the maxillary artery and maxillary division of the trigeminal nerve during the transmaxillary sinus approach to the maxillary artery. The temporal extension passes posterosuperiorly from the buccinator muscle under the zygomatic arch and lateral to the coronoid process of the mandible, separating the temporalis muscle from the arch. The superficial temporal fat pad is a separate structure, separated from the buccal fat pad's temporal extension by the deep layer of the deep temporalis fascia.

The blood supply of the buccal fat pad is from the branches of the facial artery, which lie along the most anterior portion, and the transverse facial artery, which has anastomoses with the facial artery in the region of the buccal extension. There are also contributions from the internal maxillary to the deeper portions, principally via the buccal artery. The nerve supply is from the buccal branch of the mandibular nerve prior to its distribution to the skin and the mucosa of the cheek.

The buccal fat pad may be used as free fat grafts, but more commonly it is removed incrementally to effect a change in cheek contour. This is most successful in a patient with prominent malar eminences in whom the concavity produced by fat removal will accentuate the height of the cheekbones. In a patient with flat malar arches, fat removal will lead to a gaunt look. The surgical approach may be external or intraoral. The buccal rami of the facial nerve are at risk in an external approach.

PLATYSMA

The platysma muscle is a flat muscle whose thickness varies widely among individuals; men generally have a thicker platysma than women. Inferiorly it inserts into the subcutaneous tissues of the subclavicular and acromial regions; superiorly it inserts into the chin at the commissures of the mouth and in the anterior one-third of the oblique line of the jaw. The platysma is a depressor of the lower lip innervated by the cervical branch of the facial nerve. Posteriorly and superiorly the fibers form an "S" and pass posteriorly to the angle of the mandible. Medially the fibers usually interdigitate and form an inverted "V" at the chin, 1 to 2 cm below it (most common), or at the level of the thyroid cartilage (Cardosa, 1980). Deep to the platysma are the submandibular glands, the facial nerve, and the facial artery. Laxity of the platysma with age creates "platysma banding" and can accentuate submandibular gland ptosis.

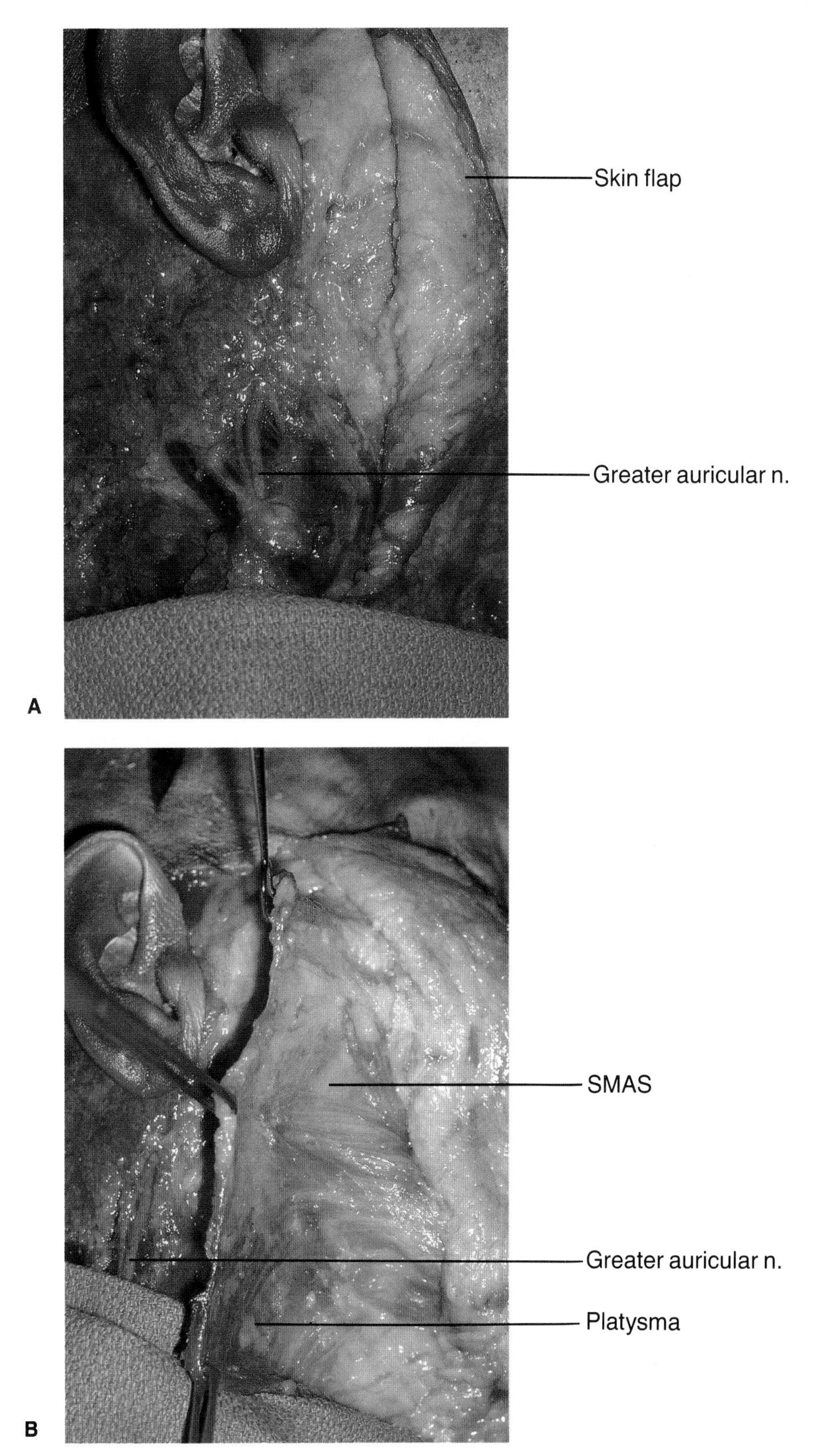

FIG. 16.1. Face-lift dissection. A: A skin flap is elevated to expose the greater auricular nerve and the underlying SMAS. **B:** The SMAS (with platysma) can be elevated as one sheet and utilized to tighten the structure of the face and the neck. Elevation anterior to the sternocleidomastoid will protect the greater auricular nerve from injury.

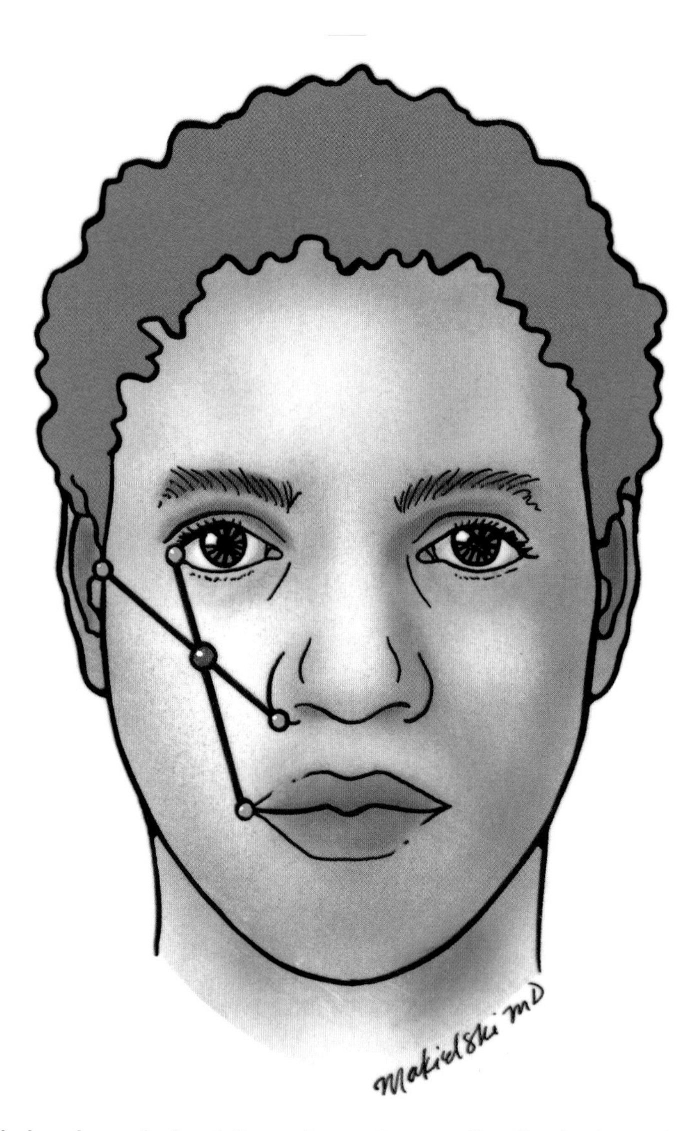

FIG. 16.2. Hinderer's analysis of the malar eminence. One line is drawn from the lateral commissure of the lip to the lateral canthus. Another line is drawn from the inferior aspect of the ala to the superior tragus. The area posterior and superior to the intersection of these lines should be the most prominent part of the malar eminence.

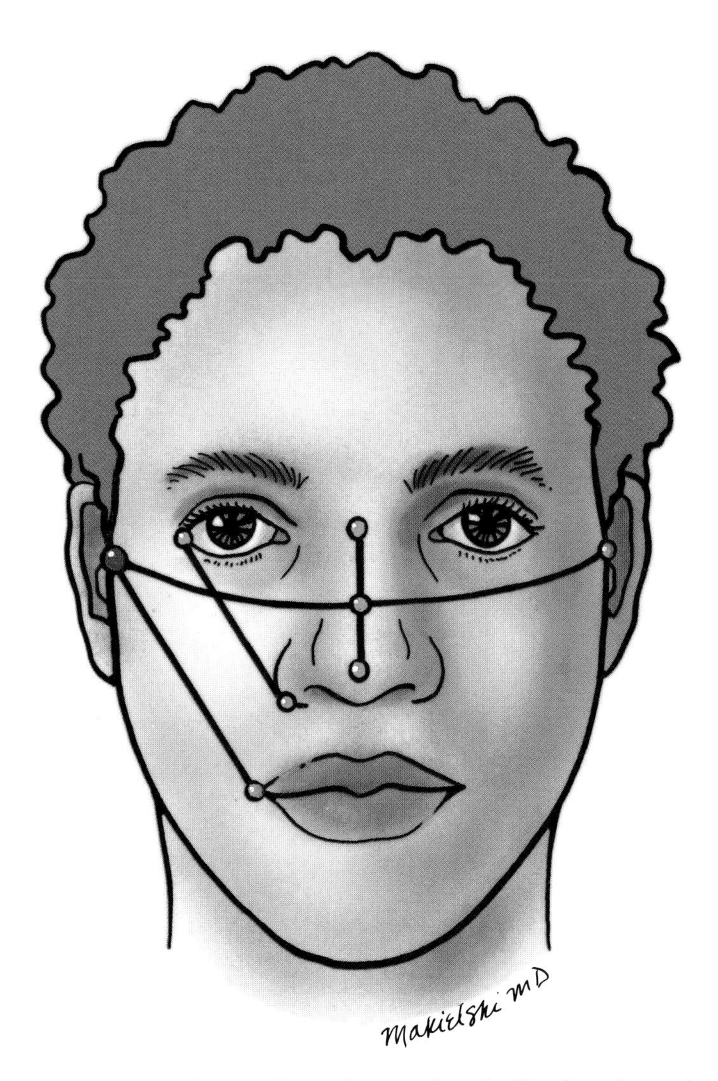

FIG. 16.3. Powell's analysis of the malar eminence. A vertical line from the nasion to the nasal tip is bisected with a curved line from the tragus of the ear; this line locates the vertical position of the malar eminence. Two other lines are drawn; the first extends from the ala to the lateral canthus and the second is drawn from the lateral commissure parallel to the first. The point where the most lateral line crosses the horizontal defines the most aesthetic location for the malar eminence.

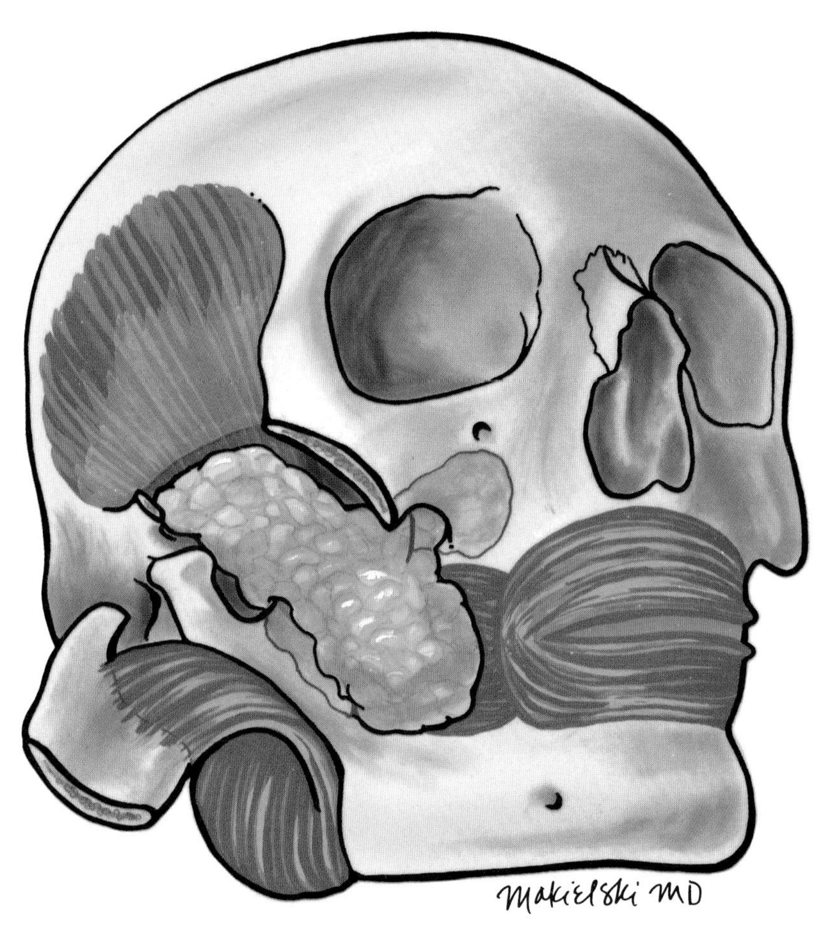

FIG. 16.4. Buccal fat pad. The zygoma and the masseter are reflected to demonstrate the buccal fat pad and its relationship to the muscles of mastication. In addition to the primary buccal extension, the pterygoid and the temporal extensions are seen. The buccal branches of the facial nerve and the parotid duct lie on the superficial surface of the buccal extension. (This relationship is shown in Fig. 5.4B.) The buccinator muscle is deep to it and must be pierced to approach the buccal fat from an intraoral approach.

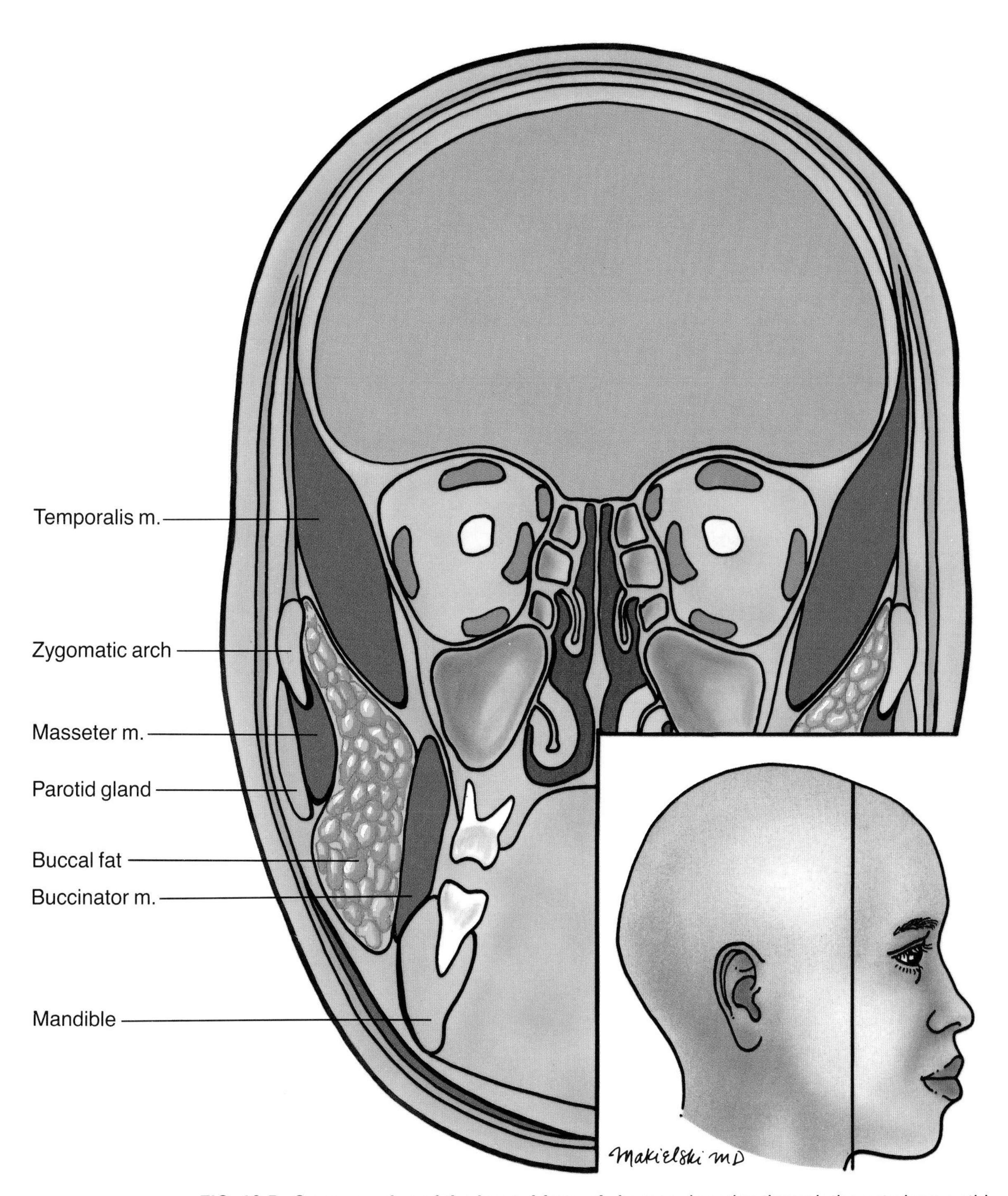

FIG. 16.5. Cross-section of the buccal fat pad. A coronal section through the anterior parotid gland demonstrates the important skeletal and muscular relationships of the buccal fat. Its primary function is to provide a gliding surface for the muscles of mastication. It thus lies between the temporalis and the masseter muscles in this plane. Its pterygoid extension surrounds the pterygoid muscles. The temporal extension of the buccal fat is separated from the superior temporal fat pad by the deep layer of the deep temporalis fascia.

CHAPTER 17

Lips and Chin

The perioral area represents one aesthetic unit, which can be divided into a number of subunits (Fig. 17.1). The lips function as an oral sphincter, a source of labial speech sounds, and a means to express emotions.

AESTHETIC PROPORTIONS OF THE LIPS

As seen in Figure 17.2, the upper lip lies slightly anterior to the lower and is about one-half the vertical height of the lower lip and the chin. The upper lip is bounded by the mesolabial fold, which descends from the alar area about 1 cm lateral to the oral commissure. The horizontal line of the chin separates the lower lip from the chin. Important landmarks of the upper lip include the philtrum column, the philtrum dimple, the white skin roll, Cupid's bow, and the tubercle. The teeth determine to a large degree the contour of the lips. On average one should see about one-third of the incisors at rest and approximately three-fourths with a smile. The horizontal width of the philtrum is about one-fourth of the width of the upper lip from commissure to commissure. Lip contour and size vary among races (Millard, 1976).

ANATOMY OF THE LIPS

The lips themselves consist of a mucosa and a submucosa, a layer of circular muscle fibers, and skin. The mucosa and the skin are firmly attached to the underlying muscle. The "wet" mucosa meets a transition area, the dry vermilion, which has a thin glandless mucosa; this vascular region is responsible for the redness of the lips.

The vascular supply of the lips is based largely on the superior and the inferior labial arteries, which are branches of the facial artery. These arteries run fairly superficially in the muscle, close to the mucosal surface, at approximately the level of the vermilion (Smith, 1961) (Fig. 17.3). The labial artery lies superficially in the muscle close to the mucosa at approximately the level of the mucocutaneous junction. Branches from the superior labial artery, such as the ascending septal branch, ascend to the base of the nose. The labial arteries must be preserved for successful use of the various cross-lip flaps.

The muscles in the perioral area are extremely well developed in humans. The sphincteric orbicularis oris arises from the region of the oral commissure. The orbicularis can be divided into a deeper pars peripheralis muscle and a more superficial pars marginalis, which underlies the vermilion. Surrounding this sphincteric muscle lies a complex of muscles. Some of these muscles elevate the upper lip (buccinator, caninus, and quadratus labii superioris); some elevate the angle of the mouth (zygomaticus major, buccinator, and caninus); some are depressors of the angle of the mouth (buccinator, depressor angularis, and risorius); others are depressors of the lower lip (mentalis, depressor labii inferioris, and platysma) (Rubin, 1977).

Melolabial (Nasolabial) Fold

The mesolabial fold is an important aesthetic landmark that separates the lip from the cheek. Its perceived depth depends on the relative fullness of the cheeks and the soft tissues of the lip plus the laxity of the cheek skin. The facial attachments to the dermis that help to create the fold (Fig. 17.4) must be recreated during facial reanimation surgery by attaching subcutaneous tissues or dermis to the underlying muscle.

THE CHIN

The chin relates visually to both the lips and the neck. Its position significantly influences the appearance of the lower face. The anterior-posterior relationship between the chin and the remainder of the profile is of practical significance. No simple measurement can define chin position exactly. Studies of aesthetic references and classical art have shown a preference for a relationship in which the lower lip is slightly posterior to the upper lip and the chin lies on a straight line connecting the two (the male chin may be somewhat more anterior). The technique of Rish is widely used. With his system, a perpendicular line is dropped from the mucocutaneous junction of the lower lip and the chin, and augmentation is considered if the chin does not reach this line. Obviously, the patient's occlusion and the functional mandibular-maxillary relationship should be considered prior to simple cosmetic chin augmentation.

When performing surgery in the anterior mandible, the major structure at risk is the mental nerve, seen in Figure 17.5.

An often overlooked component of chin position is the mentalis muscle (Zide and McCarthy, 1989). These paired midline muscles originate on the anterior mandible beneath the incisors and run inferiorly to attach to the skin of the chin. They are thus chin elevators and indirectly elevate the lip. If these muscles are cut or weakened, chin ptosis can result. The mentalis muscles, like the lip depressors, are innervated by the marginal mandibular branch of the facial nerve.

FIG. 17.1. Aesthetic subunits of the lips. The upper lip can be divided into medial and lateral subunits. The lateral subunit begins at the melolabial (nasolabial) fold and ends at the philtrum column. Superolaterally, the small triangle at the base of the ala should be included in the aesthetic subunit. Inferiorly, both lateral and medial subunits follow the vermilion border. Reconstruction of a complete subunit places incisions at the boundaries where they are least noticed.

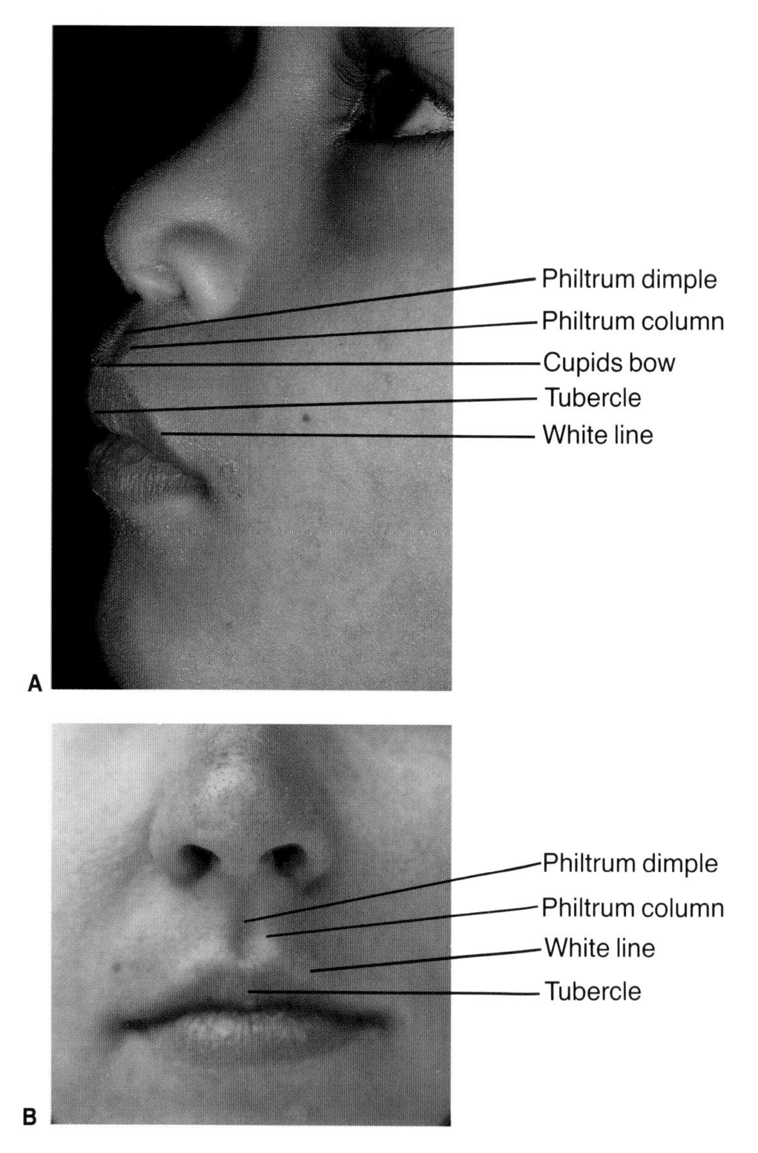

FIG. 17.2. Landmarks of the lips. The philtrum, the philtral crest, Cupid's bow, the vermilion, the white line, and the tubercle are shown. The upper lip lies somewhat more anteriorly than the lower. **A:** Lateral view. **B:** Frontal view.

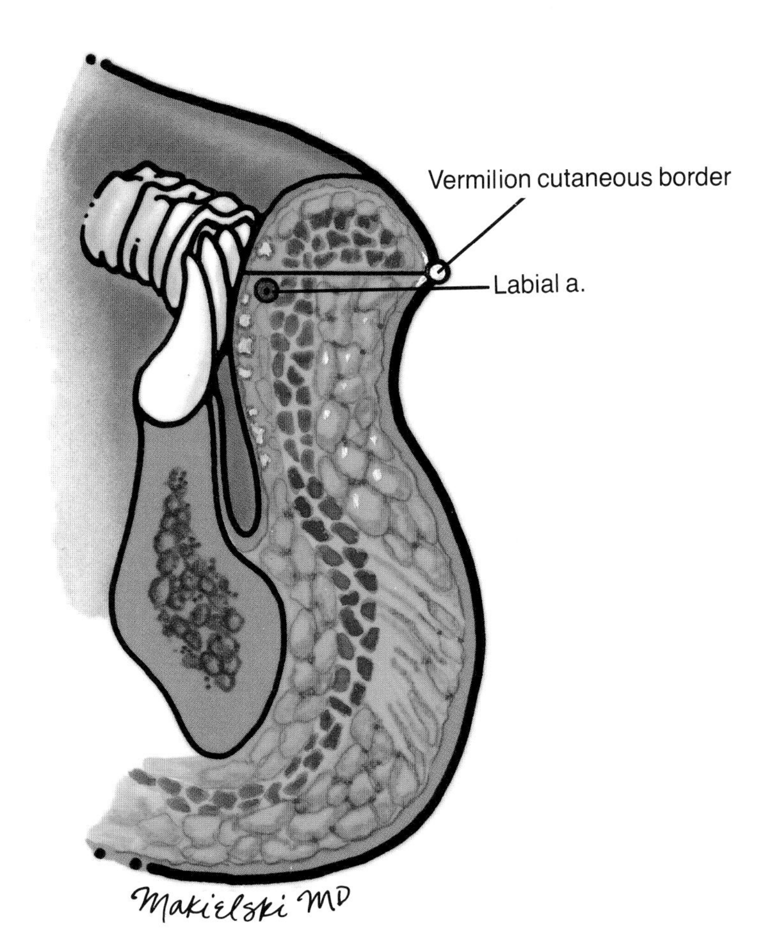

FIG. 17.3. Cross-section of the lips. The labial artery runs in the orbicularis oris muscle, deep to the mucosa, at approximately the level of the vermilion border.

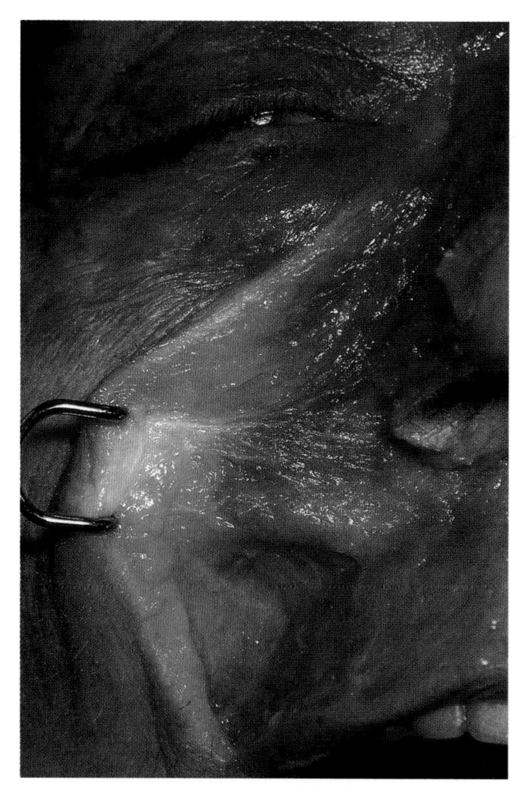

FIG. 17.4. Melolabial (nasolabial) crease. The SMAS is shown inserting into the dermis. The contraction of the attached SMAS and the facial musculature creates the melolabial crease.

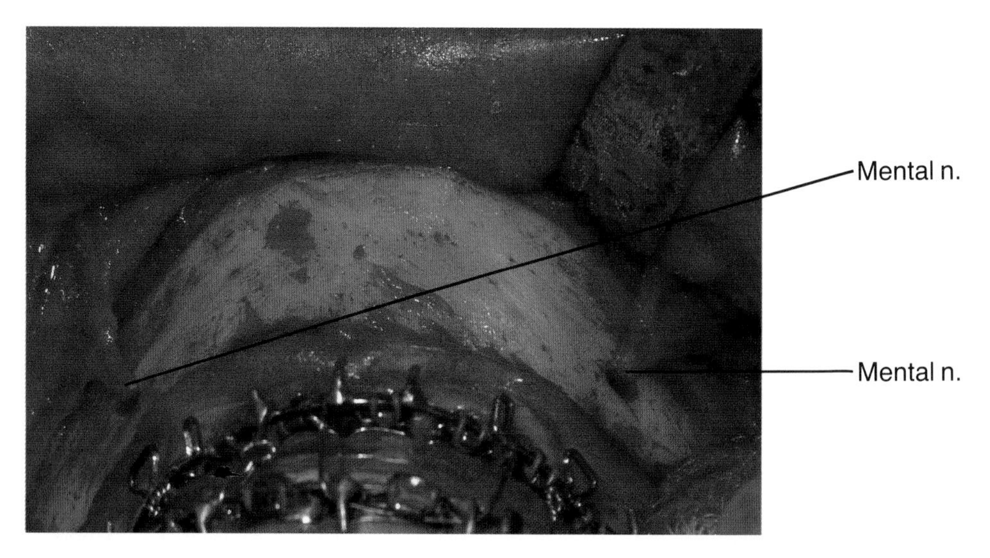

FIG. 17.5. Mental nerves. The mental nerves are exposed intraorally. These nerves provide sensation to the lower lip and lie approximately in the midpupillary line. (Courtesy of Gary Feldman, M.D., D.D.S.)

Selected Readings

FACIAL CONTOUR ANALYSIS

1. Bernstein L: Esthetic anatomy of the nose. *Laryngoscope* 1971;82:1323–1330.
2. Broadbent BH Sr, Broadbent BH Jr, Golden WH: *Bolton standards of dentofacial developmental growth.* St. Louis: CV Mosby, 1975.
3. Brown JB, McDowell F: *Plastic surgery of the nose.* St. Louis: CV Mosby, 1951:30–34.
4. Gallagher DM, Bell WH, Storum KA: Soft tissue changes associated with advancement genioplasty performed concomitantly with superior repositioning of the maxilla. *J Oral Maxillofac Surg* 1984;42:238–242.
5. Gonzalez-Ulloa M: Quantitative principles in cosmetic surgery of the face (profileplasty). *Plast Reconstr Surg* 1961;29:186–198.
6. Holdaway R: A soft-tissue cephalometric analysis and its use in orthodontic treatment planning, part I. *Am J Orthod* 1983;84:1–28.
7. Holdaway R: A soft-tissue cephalometric analysis and its use in orthodontic treatment planning, part II. *Am J Orthod* 1984;85:279–293.
8. Larrabee WF Jr, Sidles J, Sutton D: Facial analysis. *Laryngoscope* 1988;98:1273–1275.
9. Legan H, Burstone C: Soft-tissue cephalometric analysis for orthognathic surgery. *J Oral Surg* 1980;38:744–751.
10. Peck H, Peck S: A concept of facial esthetics. *Angle Orthod* 1970;40:284–317.
11. Powell N, Humphreys B: *Proportions of the aesthetic face.* New York: Thieme-Stratton, 1984.
12. Riola ML, Moyers RE, McNamara JA, Hunter WS: *An atlas of craniofacial growth.* Ann Arbor: University of Michigan, 1974.
13. Simons RL: Adjunctive measures in rhinoplasty. *Otolaryngol Clin North Am* 1975;8:717–742.
14. Steiner CC: Cephalometrics in clinical practice. *Angle Orthod* 1959;29:8–29.
15. Zide B, Grayson B, McCarthy J: Cephalometric analysis, part I. *Plast Reconstr Surg* 1981;68:816–823.
16. Zide B, Grayson B, McCarthy J: Cephalometric analysis for upper and lower midface surgery, part II. *Plast Reconstr Surg* 1981;68:961–968.
17. Zide B, Grayson B, McCarthy J: Cephalometric analysis for mandibular surgery, part III. *Plast Reconstr Surg* 1982;69:155–164.

VARIATIONS IN FACIAL ANATOMY WITH RACE, SEX, AND AGE

1. Alexander RC, Hitchcock HP: Cephalometric standards for American Negro children. *Am J Orthod* 1978;74:298–309.

2. Daly CH, Odland GF: Age-related changes in the mechanical properties of human skin. *J Invest Dermatol* 1979;73(1):84–87.
3. Dedo DD: A preoperative classification of the neck for cervicofacial rhytidectomy. *Laryngoscope* 1988;90:1894–1896.
4. Drummond RA: A determination of cephalometric standards for the Negro race. *Am J Orthod* 1968;54:670–682.
5. Garcia CJ: Cephalometric evaluation of Mexican Americans using the Downs and Steiner analyses. *Am J Orthod* 1975;68:67–74.
6. Gilchrest BA, Szabo G, Flynn E, et al.: Chronologic and actinically induced aging in human facial skin. *J Invest Dermatol* 1983;80(Suppl 6):81S–85S.
7. Gonzalez-Ulloa M, Flores ES: Senility of the face: Basic study to understand its causes and effects. *Plast Reconstr Surg* 1965;36(Aug):239–246.
8. Gonzalez-Ulloa M, Simonin F, Flores ES: The anatomy of the aging face. *Trans Fifth Int Cong Plast Reconstr Surg* 1971;1059–1065.
9. Guo MK: Cephalometric standards of Steiner Analysis established on Chinese children. *J Formosan Med Assoc* 1971;70(2):97–102.
10. Hinderer KH: *Fundamentals of anatomy and surgery of the nose.* Birmingham, AL: Aesculapius Publishing, 1971:54.
11. Krmpotic-Nemanic J, Kostovic I, Rudan P, et al.: Morphological and histological changes responsible for the droop of the nasal tip in advanced age. *Acta Otolaryngol (Stockh)* 1971;71:278–281.
12. Larrabee WF, Sutton D, Carlisle KS: A histologic and mechanical study of the aging skin. In: Ward PW, Berman WE, eds. *Proceedings of the 4th international symposium on plastic and reconstructive surgery of the head and neck.* St. Louis: CV Mosby, 1984:293–296.
13. Lines PA, Lines RL, Lines CA: Profilemetrics and facial esthetics. *Am J Orthod* 1978;73:648–657.
14. Miura F, Inoue N, Suzuki K: Cephalometric standards for Japanese according to the Steiner analysis. *Am J Orthod Dentofacial Orthop* 1965;51:288–295.
15. Montagna W, Carlisle KS: Structural changes in aging human skin. *J Invest Dermatol* 1979;73(1):47–53.
16. Taylor WH, Hitchcock HP: The Alabama analysis. *Am J Orthod* 1966;52:245–265.

HARD-TISSUE FOUNDATION

1. Lowrey GH: *Growth and development of children,* 8th ed. Chicago: Yearbook Medical Publishers, 1986.
2. Manson P: Facial injuries. In: McCarthy JG, ed. *Plastic surgery, vol 2.* Philadelphia: WB Saunders, 1990:867–1141.
3. McVay CB: *Surgical anatomy.* Philadelphia: WB Saunders, 1984.
4. Pensler J, McCarthy JG: The calvarial donor site: An anatomic study in cadavers. *Plast Reconstr Surg* 1984;75:648–651.
5. Smith JD, Abramson M: Membranous versus endochondral bone autografts. *Arch Otolaryngol* 1974;99:203–205.
6. Toriumi DM, Kotler HS, Luxenberg DP, Holtrip ME, Wang EA: Mandibular reconstruction with a recombinate bone-inducing factor. Functional, histologic, and biomechanical evaluation. *Arch Otolaryngol Head Neck Surg* 1991;117(10):1101–1112.
7. Toriumi DM, East CA, and Larrabee WF Jr: Osteoinductive biomaterial for medical implantation. *J Long-Term Effects of Med Implants* 1991;1:53–77.
8. Zins J, Whitaker L: Membranous versus endochondral bone: Implications for craniofacial reconstruction. *Plast Reconstr Surg* 1983;72:778–785.

SKIN AND SOFT TISSUE

1. Borges AF: *Elective incisions and scar revision.* Boston: Little, Brown, 1973.
2. Burget GC, Menick FJ: Nasal reconstruction: Seeking a fourth dimension. *Plast Reconstr Surg* 1986;78:145–157.
3. Burget GC, Menick FJ: The subunit principal in nasal reconstruction. *Plast Reconstr Surg* 1985;76:239–247.

4. Gonzalez-Ulloa M, Costillo A, Stevens E, Fuertes GA, Leovilli F, Ubaldo F: Preliminary study of the total restoration of the facial skin. *Plast Reconstr Surg* 1954;13:151–161.
5. Larrabee WF Jr: A finite element model of skin deformation I–III. *Laryngoscope* 1986;96:399–419.
6. Larrabee WF Jr, Sutton D: Variation of skin stress-strain curves with undermining. *Surg Forum* 1981;32:553–555.
7. Larrabee WF Jr, Holloway GA Jr, Sutton D: Wound tension and blood flow in skin flaps. *Ann Otol Rhinol Laryngol* 1984;93:112–115.

SUPERFICIAL MUSCULOAPONEUROTIC SYSTEM

1. Abul-Hassan HS, Ascher G, Acland RD: Surgical anatomy and blood supply of the fascial layers of the temporal region. *Plast Reconstr Surg* 1986;77:17–24.
2. Dzubow LM: *Facial flaps biomechanics and regional applications.* CT: Norwalk, Appleton and Lange, 1990.
3. Jost G, Levet Y: Parotid fascia and face-lifting: A critical evaluation of the SMAS concept. *Plast Reconstr Surg* 1984;74:42–51.
4. Letourneau A, Daniel R: The superficial musculoaponeurotic system of the nose. *Plast Reconstr Surg* 1988;82:48–55.
5. Mitz V, Peyronie M: The superficial musculo-aponeurotic system (SMAS) in the parotid and cheek area. *Plast Reconstr Surg* 1976;58:80–88.
6. Pensler J, Ward J, Parry S: The superficial musculoaponeurotic system in the upper lip: An anatomic study in cadavers. *Plast Reconstr Surg* 1985;75:488–492.

FACIAL MUSCULATURE

1. Correia PdeC, Zani R: Masseter muscle rotation in the treatment of inferior facial paralysis. *Plast Reconstr Surg* 1973;52:370–373.
2. Conley JJ: *Salivary glands and the facial nerve.* New York: Grune and Stratton, 1975.
3. Faigin G: *The artist's complete guide to facial expression.* New York: Watson-Guptill Publications, 1990.
4. Lightoller GHS: Facial muscles. *J Anat* 1925;60:1–85.
5. Rubin LR: Anatomy of a smile. *Plast Reconstr Surg* 1974;53:384–387.

FACIAL NERVE

1. Baker DC, Conley J: Avoiding facial nerve injuries in rhytidectomy. *Plast Reconstr Surg* 1979;64:781–795.
2. Bernstein L, Nelson RH: Surgical anatomy of the extraparotid distribution of the facial nerve. *Arch Otolaryngol* 1984;110:177–183.
3. Correia PdeC, Zani R: Surgical anatomy of the facial nerve as related to ancillary operations in rhytidoplasty. *Plast Reconstr Surg* 1973;52:549–552.
4. Davis BA, Anson BJ, Budinger JM, et al: Surgical anatomy of the facial nerve and the parotid gland based upon a study of 350 cervico-facial halves. *Surg Gynecol Obstet* 1956;102:385–412.
5. Dingman RO, Grabb WC: Surgical anatomy of the mandibular ramus of the facial nerve based on the dissection of 100 facial halves. *Plast Reconstr Surg* 1962;29:266–272.
6. Liebman EP, Webster RC, Berger AS, Della Vecchia M: The frontalis nerve in the temporal brow lift. *Arch Otolaryngol* 1982;108:232–235.
7. Liebman EP, Webster RC, Gaul JR, Griffin T: The marginal mandibular nerve in rhytidectomy and liposuction surgery. *Arch Otolaryngol Head Neck Surg* 1988;104:179–181.
8. Rudolph R: Depth of the facial nerve in face lift dissections. *Plast Reconstr Surg* 1990;85:537–544.
9. Stuzin JM, Wegstrom L, Kawamoto HK, Wolfe SA: Anatomy of the frontal branch of the facial nerve: The significance of the temporal fat pad. *Plast Reconstr Surg* 1989;83:265–271.
10. Tabb HG, Tannehill JF: The tympanomastoid fissure: A reliable approach to the facial nerve in parotid surgery. *South Med J* 1973;66:1273–1276.

FACIAL SENSORY INNERVATION

1. Grant JCB: *An atlas of anatomy*, 6th ed. Baltimore: Williams and Wilkins, 1972.
2. Woelfel JB: *Dental anatomy*. Philadelphia: Lea and Febiger, 1990.

VASCULAR PATTERNS OF THE FACE

1. Daniel RK, Williams HB: The free transfer of skin flaps by microvascular anastomoses: An experimental study and reappraisal. *Plast Reconstr Surg* 1973;52:16–31.
2. Daniel RK, Kerrigan CL: Skin flaps: An anatomical and hemodynamical approach. *Clin Plast Surg* 1979;6:181–200.
3. Daniel RK, Kerrigan CL: Principles and physiology of skin flap surgery. In: McCarthy JG, ed. *Plastic surgery, vol 1*. Philadelphia: WB Saunders, 1990:275–328.
4. Hagan WE, Walker LB: The nasolabial musculocutaneous flap: Clinical and anatomic correlations. *Laryngoscope* 1988;98:341–346.
5. Larrabee WF Jr: Design of local skin flaps. *Otolaryngol Clin North Am* 1990;23:899–923.
6. Larrabee WF Jr: A discussion of the use of bilobed flaps for repair of large temple defects. *Arch Otolaryngol–Head and Neck Surg* 1992;118:983–984.
7. Manchot C: *The cutaneous arteries of the human body*. New York: Springer-Verlag, 1983.
8. Mathes SJ, Nahai F: Classification of the vascular anatomy of muscles: Experimental and clinical correlation. *Plast Reconstr Surg* 1981;67:177–187.
9. Salmon M: *Arteries of the skin*. New York: Churchill-Livingstone, 1988.
10. Taylor GI, Daniel RK: The anatomy of several free flap donor sites. *Plast Reconstr Surg* 1975;56:243–253.
11. Taylor GI, Palmer JH: The vascular territories (angiosomes) of the body: Experimental study and clinical applications. *Br J Plast Surg* 1987;40:113–141.
12. Taylor GI, Palmer JH, Mchanamny D: The vascular territories of the body (angiosomes) and their clinical applications. In: McCarthy JG, ed. *Plastic surgery, vol 1*, Philadelphia: WB Saunders, 1990:329–378.
13. Tolhurst DE, Haeseker B, Zeeman RJ: The development of the fasciocutaneous flap and its clinical applications. *Plast Reconstr Surg* 1983;7:597–605.

LYMPHATICS OF THE FACE

1. Cassisi NJ, Dickerson DR, Million RR: Squamous cell carcinoma of the skin metastatic to the parotid nodes. *Arch Otolaryngol Head Neck Surg* 1978;104:336–339.
2. Goepfert H, Jesse RH, Ballantyne AJ: Posterolateral neck dissection. *Arch Otolaryngol* 1980;106:618–620.
3. Graham JW: Cancer in the parotid lymph nodes. *Med J Aust* 1965;2:8–12.
4. Hollinshead WH: *Textbook of anatomy*, 3rd ed. Philadelphia: Harper and Row, 1974:824–825.
5. Lingeman RE, Schellhamer RH: Surgical management of tumors of the neck. In: Thawley SE, Panje WR, eds. *Comprehensive management of head and neck tumors*. Philadelphia: WB Saunders, 1987:1325–1350.
6. Shah JP, Strong E, Spiro RH, Vikram B: Surgical grand rounds, neck dissection: Current status and future possibilities. *Clin Bull* 1981;11:25–33.
7. Som P: Lymph nodes of the neck. *Radiology* 1987;165:593–600.

HAIR AND SCALP

1. Baden HP: *Diseases of the hair and nails*. Chicago: Yearbook Medical Publishers, 1987.
2. Norwood OT: Male pattern baldness: classification and incidence. *South Med J* 1975;68:1359–1365.

3. Norwood OT, Shiel RC: *Hair transplant surgery,* 1st ed. Springfield, IL: Charles C Thomas, 1973:6.
4. Tolhurst DE, et al.: The surgical anatomy of the scalp. *Plast Reconstr Surg* 1991;87:603–612.

FOREHEAD AND BROW

1. Brennan HG: The forehead lift. *Otolaryngol Clin North Am* 1980;13:209–233.
2. Castañares S: Forehead wrinkles, glabellar frown, and ptosis of the eyebrow. *Plast Reconstr Surg* 1964;34:406–413.

EYELIDS, ANTERIOR ORBIT, AND LACRIMAL SYSTEM

1. Beard C, Quickert MN: *Anatomy of the orbit,* 2nd ed. Birmingham, AL: Aesculapius Publishing, 1977.
2. Callahan M, Beard C: *Beard's ptosis,* 4th ed. Birmingham, AL: Aesculapius Publishing, 1990.
3. Hornblass A (ed.): *Oculoplastic, orbital, and reconstructive surgery, vol I: eyelids.* Baltimore: Williams and Wilkins, 1988.
4. Liu D, Hsu WM: Oriental eyelids. *Ophthalmic Plast Reconstr Surg* 1986;2:59–64.
5. Reeh MJ, Wobig JL, Wirtschaffer S: *Ophthalmic anatomy.* American Academy of Ophthalmology, 1981.
6. Zide BM, Jelks GW: *Surgical anatomy of the orbit.* New York: Raven Press, 1985.

NOSE

1. Anderson JR: A reasoned approach to nasal base surgery. *Arch Otolaryngol* 1984;110:349–358.
2. Bernstein L: Surgical anatomy in rhinoplasty. *Otolaryngol Clin North Am* 1975;8:549–558.
3. Burget GC, Menick FJ: The subunit principles in nasal reconstruction. *Plast Reconstr Surg* 1985;76:239–247.
4. Crumley RJ, Lancer M: Quantitative analysis of nasal tip projection. *Laryngoscope* 1988;98:202–208.
5. Griesman B: Muscles and cartilages of the nose from the standpoint of a typical rhinoplasty. *Arch Otolaryngol* 1944;39:334–341.
6. Janeke JB, Wright WK: Studies on support to the nasal tip. *Arch Otolaryngol* 1971;93:458–464.
7. Krmpotic-Nemanic J, Kostovic I, Rudan P, et al.: Morphological and histological changes responsible for the droop of the nasal tip in advanced age. *Acta Otolaryngol (Stockh)* 1971;72(2):278–281.
8. Powell N, Humphreys B: *Proportions of the aesthetic face.* New York: Thieme-Stratton, 1984.
9. Tardy ME, Brown RJ: *Surgical anatomy of the nose.* New York: Raven Press, 1990.

EARS

1. Adamson JE, Horton CE, Crawford HH: The growth pattern of the external ear. *Plast Reconstr Surg* 1965;36:466–469.
2. Anson BJ, McVay CB: *Surgical anatomy,* 6th ed. Philadelphia: WB Saunders, 1984:110.
3. Bardach J: Surgery for congenital and acquired malformations of the auricle. In: Cummings CW, Fredrickson JM, Hawker LA, Krause CJ, Schuller DE, eds. *Otolaryngology: head and neck surgery.* St. Louis: CV Mosby, 1986:2861–2876.
4. Brent B: Reconstruction of the auricle. In: McCarthy JG, ed. *Plastic surgery.* Philadelphia: WB Saunders, 1990:2094.

5. Hollinshead WH: *Anatomy for surgeons, vol 1, the head and neck.* New York: Harper and Brothers, 1961:166.
6. Larrabee WF Jr, Kibblewhite D, Adams J: Otoplasty in Pediatric Facial Plastic Surgery. In: Bumstead R, Smith J, eds. New York: Raven Press, 1993, *in press.*
7. Levine H: Auricular and periauricular cutaneous carcinomas. In: Thawley SE, Panje WR, eds. *Comprehensive management of head and neck tumors.* Philadelphia: WB Saunders, 1987:195–206.
8. Maniglia AJ, Maniglia JV: Congenital lop ear deformity. *Otolaryngol Clin North Am* 1981;14:83–93.
9. Polyak SL, McHugh G: In: McKenna TH, ed. *The human ear.* Sonotone Corporation: New York, 1946:53.
10. Rogers BO: Microtic, lop, cup and protruding ears: Four directly inheritable deformities? *Plast Reconstr Surg* 1968;41:208–231.
11. Webster RC, Smith RC: Otoplasty for prominent ears. In: Goldwin RM, ed. *Long-term results in plastic and reconstructive surgery.* Boston: Little, Brown, 1980:146.
12. Wright WK: Otoplasty goals and principles. *Arch Otolaryngol Head Neck Surg* 1970;92:568–572.

CHEEKS AND NECK

1. Cardosa C: The anatomy of the platysma muscle. *Plast Reconstr Surg* 1980;66:680–683.
2. McKinney P, Gottlieb J: The relationship of the greater auricular nerve to the SMAS. *Ann Plast Surg* 1985;14:310–314.

LIPS AND CHIN

1. Burget GC, Menick FJ: Aesthetic restoration of one-half of the upper lip. *Plast Reconstr Surg* 1986;78:583–593.
2. Millard DR: *Cleft craft, vol 1.* Boston: Little, Brown, 1976.
3. Rubin LR: *Reanimation of the paralyzed face.* St. Louis: CV Mosby, 1977.
4. Smith JW: Clinical experience with the vermilion bordered lip flap. *Plast Reconstr Surg* 1961;27:527–543.
5. Zide BM, McCarthy J: The mentalis muscle: An essential component of chin and lower lip position. *Plast Reconstr Surg* 1989;83:413–420.

Subject Index

Page numbers appearing in italic refer to figures in the text.